A FISHKEEPER'S GUIDE TO

Xiphophorus variatus (*Variatus Platy*)

Rasbora heteromorpha (*Harlequin Fish*)

A splendid introduction to the care and breeding of 60 freshwater tropical fishes for the community tank

Dick Mills

TetraPress

No. 16062

A Salamander Book

Published in the USA by Tetra Press,
201 Tabor Road,
Morris Plains,
N.J. 07950

ISBN 3-923880-52-9

Barbus conchonius (*Rosy Barb*)

Credits

Editor: Geoff Rogers Designer: Mark Holt
Colour reproductions:
Bantam Litho Ltd.
Rodney Howe Ltd.
Filmset: SX Composing Ltd.
Printed by Proost International Book Production, Turnhout, Belgium.

Author

The author, Dick Mills, has been keeping fishes for over 30 years, during which time he has written many articles for aquatic hobby magazines as well as 12 books. A member of his local aquarist society, for the past 20 years he has also been a Council member of the Federation of British Aquatic Societies, for which he regularly lectures and produces a quarterly News Bulletin. By profession, he composes electronic music and special sound sequences for television and radio programmes – a complete contrast to fishkeeping, the quietest of hobbies.

Consultant

Fascinated by fishkeeping from early childhood, Dr. Neville Carrington devised an internationally known liquid food for young fishes while studying for a pharmacy degree. After obtaining his Doctorate in Pharmaceutical Engineering Science and a period in industry, Dr. Carrington now pursues his life-long interest in developing equipment and chemical products for the aquarium world.

Contents

Introduction

Naturally the newcomer to fishkeeping will be excited and anxious to get going with his first tank of fishes. He may be bewildered by the choice available and, again, because of his newness to the hobby, may well start off by choosing unwisely.

The attraction of keeping tropical fishes is that so many different species are available from all parts of the tropical fresh waters of the world, including a few species from slightly cooler waters that nevertheless can be included with their warmer-water relatives. At the beginning of the fishkeeping interest, the newcomer will have no idea how one group of fishes can differ from another, or whether they will be compatible in the confines of the aquarium; and there is no opportunity for any fish to swim away from trouble.

The beginning aquarist tends to choose a 'bit of everything' with which to stock the first aquarium; only later, with acquired knowledge, does he start to specialize. The first collection of fishes is generally described as a 'community' although there is no reason at all why a collection of differing species should automatically get on well together just because a fishkeeper has chosen them as his first pets in the newly furnished aquarium.

Within the pages of this book, the phrase 'community fishes' will be used to describe those fishes that will all tolerate similar aquarium conditions such as water temperature and food, and have either a peaceable disposition or a healthy disregard of each other – no one has yet discovered which of these conditions is true.

Again, the newcomer might consider keeping only related or single species in his aquarium; this would be a waste of effort if the full potential of the fish-holding capacity of the aquarium were ignored. Generally, however, most fishkeepers want to stock the aquarium to its capacity and this means selecting fishes that naturally occupy all levels of water in the aquarium, which belong to different species.

This guide has been structured to present the fishes to you in three categories: surface swimmers, mid-water fishes and bottom-dwelling species. None of the species shown is obligatory; the choice is left open to your own particular tastes and to what is commercially available in your area. However, within the species described there are fishes that have been found to be compatible when kept together and these will get you off to a sound start. The first part is most important; please do not turn immediately to the fish illustrations, however tempting they appear. Find out the basic needs of the species and how to care for them first, then you can go ahead and make your choice of fishes. In this way, you may well be confident enough to make use of the third part of the guide, which describes the breeding of the fishes. After this, it is all up to you, as you continuously reconsider what your idea of a community fish collection will be. That is the beauty of fishkeeping; there is always something new to try, with the promise of new satisfactions and achievements. Who could ask for a better hobby?

Tank size

When choosing the size of the tank there are certain factors to be considered. Obviously, the number of fishes that can be kept in it will be uppermost in most fishkeepers' minds; there is also the question of how the overall dimensions of the tank fit in where you intend it to be sited. Again, some tank shapes are more pleasing than others and lend themselves better to aquatic 'landscaping' and furnishing when setting up the tank.

It is suggested that the minimum length of tank should be 60cm (24in) with a front to rear dimension of 30cm (12in) and a depth of 30cm (12in), but a better aquatic picture is provided with a tank depth of 38cm (15in).

Although tanks can be made to suit any required dimensions and to fit exactly into any awkward recess, it should be borne in mind that the tank will be the living quarters of our fishes; and although the dimensions of the tank may suit *our* requirements, it does not always follow that the fishes will be happy with this arrangement. For reasons that will be explained in a moment, the number of fishes that can be accommodated within any tank is governed by the length and breadth, and the depth of the tank affects only the water capacity.

Through their gills fishes absorb oxygen that is dissolved in water, and they exhale carbon dioxide. As the oxygen is depleted it is replenished via the water surface, the interface between water and air. Similarly,

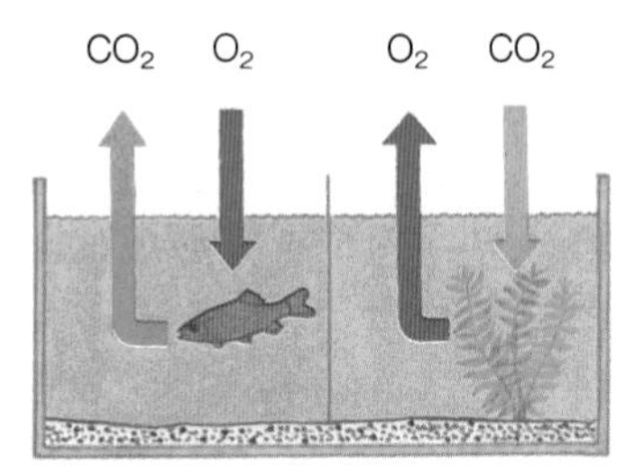

Above: *During the hours of 'tanklight' the fishes' and plants' respiratory needs and actions complement each other perfectly in the aquarium.*

carbon dioxide can be expelled only at the water surface, although we shall see later how aquarium plants can reduce the level of carbon dioxide in the water. So, to allow the tank to accommodate its full quota of fish, without discomfort and the danger of asphyxiation, a large water surface is required. Reference to the diagram will show you that although the two tanks hold the same volume of water, the one with the largest water surface can hold the largest number of fishes. The general rule is to allow $75cm^2$ ($12in^2$) for every 2.5cm (1in) of fish body. (When measuring a fish do not include the tail.) For example a tank 60cm × 30cm (24in × 12in) will hold 24 fishes each 2.5cm long. This figure holds true no matter how deep the tank is, so in the case of our 60cm (24in) tank you may as well opt for the deeper 38cm (15in) dimension, and give your fishes more room to swim in

Below: *In two tanks of identical volume a higher number of fishes can be kept in the tank with the larger water surface area, shown at right.*

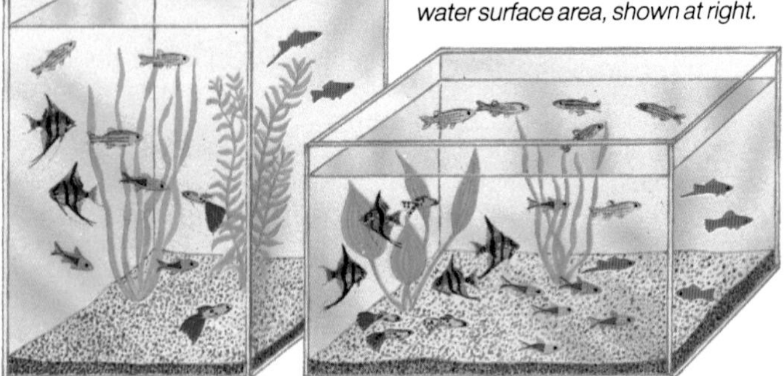

Above: *Careful planning and choice of species ensure that a varied, well-stocked aquarium makes use of all the space available, with fishes swimming at all levels. The three levels – (1) top (2) mid-water (3) bottom – form the divisions of the fish section later in this guide.*

and yourself more room in which to create your underwater world.

Another reason for choosing a 60cm (24in) tank is that within the confines of the aquarium the water conditions will fluctuate, particularly in small tanks. With this tank size (capacity 54-68 litres, 12-15 gallons) such changes will be more gradual and less likely to affect the fishes adversely. Similarly, temperature changes will occur more gradually.

A 'balanced' aquarium

Having selected a suitably sized tank, you can now fill it with fishes. The aquarium should present to the viewer a complete picture of an underwater world; fishes should be chosen that will naturally inhabit all water levels in the aquarium. Surface-swimming fishes tend to be fast swimmers, either to escape predators or to catch their own food, usually insects floating on the water surface. Mid-water swimmers can be more slow-moving; for example, the Angelfish (*Pterophyllum* sp.) inhabits reedbeds and is quite a sedate fish, as are members of the Gourami family. Bottom-dwelling fishes are another characteristic group; many are nocturnal by nature, hiding away among plants and rocks by day (or tank-lit times). They are often relegated to the scavenger category by ignorant fishkeepers who do not realize that these fishes have a lifestyle of their own and deserve to be treated accordingly.

There are no set rules for the proportions of surface swimmers to mid-water fishes to bottom dwellers; obviously, species from one group will stray into the other two strata during their daily lives and not even the most inexperienced fishkeeper would expect it to be otherwise. Generally, the mid-water fishes outnumber the other two groups in the aquarium just as they do in nature. Some aquarists deliberately specialize in fishes from only one group, but that is a choice you can make as you progress in your own fishkeeping adventures.

Heating and lighting

As we are concerned with a mixed collection of fishes we have to steer a middle course and hope that our conditions will suit most of the fishes for most of the time.

Fishes coming from the tropical areas require a reasonably constant temperature of around 24°C (75°F). Again, because of our chosen cross-section of species, some will be used to wider temperature variations than others; a shallow tropical pool will vary more rapidly in temperature during the day than a large deep river. At night, the temperature in the wild may well drop more degrees that we expect. The only time that very accurate maintenance of the water temperature is essential, however, is when attempts are made to persuade your fishes to breed, and even then it may be necessary to be very exact with only one or two species.

Heating the aquarium

The modern aquarium is heated by miniature immersion heaters controlled by a built-in thermostat. The thermostat will switch off the electricity supply to the heater when the desired temperature has been reached, and will switch on the supply again when the temperature has fallen a degree or two. With a chosen tank size of 60cm (24in) or more, the body of water has a large enough thermal capacity not to cool down too quickly when the heating is switched off (either by the thermostat, or by accident or a power failure). It follows, too, that the aquarium heater is not on all the time, and horrendously high electricity bills are avoided.

As a guide, 10 watts of 'heat' (aquarium heaters are rated in watts) should be allowed for every 4.5 litres (1 gallon) of water. Our 60cm (24in) aquarium holding 54-68 litres (12-15 gallons) depending on its depth, will require 120-150 watts of heat – ie, a 125 or 150 watt heater will be adequate. Heater-thermostat units

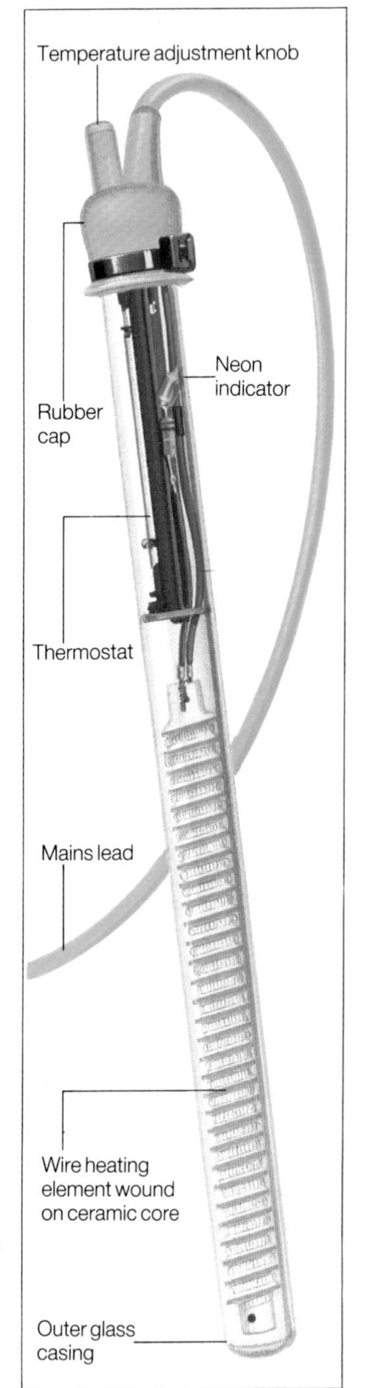

Right: *This heater-thermostat unit is electro-mechanically operated and magnetic algae scrapers should not be parked near it. More recent models have solid-state circuitry.*

are sold in standard sizes; choose one with a slightly higher wattage than your theoretical calculated size if an exact match is not possible.

Tanks more than 90cm (36in) long are best warmed by two heaters (together totalling the required heat wattage); the reason for this is that two heaters can provide a more even spread of heat throughout the tank.

Heat losses through the glass sides of the tank can be minimized by covering the outside of the sides and back of the tank with sheets of expanded polystyrene.

Lighting the aquarium

Even if fishes come from the tropics where the sun beats down from practically overhead for 12 hours of each day, not all are sun-lovers. Many seek the shade during the day, even to the extent of following a patch of shade around a pool as the sun progresses across the sky.

The tank is easily lit from above, by installing the lighting equipment in the tank's cover, usually referred to as a hood or reflector.

If you are using tungsten lighting (filament lamps), you will need around 40 watts per $900cm^2$ ($1ft^2$) of water surface area (slightly more for a deep tank). However, most aquarists find that tungsten lighting produces too much heat (shortening lamp life too) and prefer to use fluorescent lighting, which is both cooler and costs less to run, needing only 10 watts per $900cm^2$. The necessary starting equipment for fluorescent tubes can be either accommodated in the tank hood or situated remotely. For safety reasons, place a clear plastic or glass sheet between the water and the lamps.

Above: *This light hood fitted with tungsten lamps has a hinged front to provide easy access to the tank for feeding and replacing the lamps.*

Below: *Various lamp combinations for aquarium use.* Top: *Tungsten lamps.* Centre: *Tungsten and fluorescent lighting.* Bottom: *Two different 'colours' of fluorescents.*

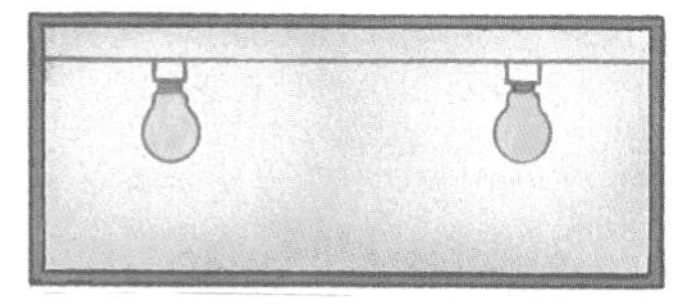

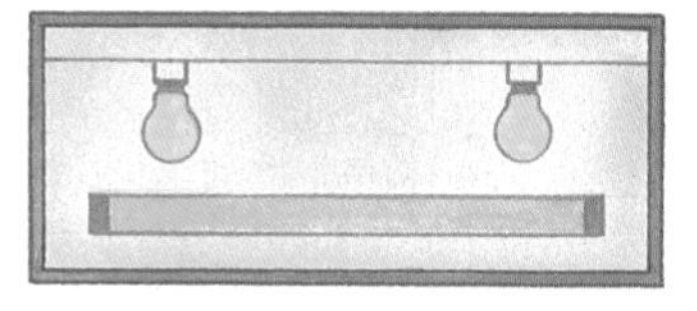

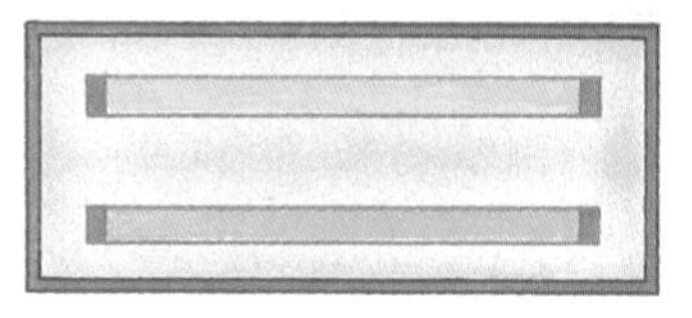

The tank should be lit for 12-15 hours each day, the duration adjusted to suit your fish-viewing requirements and to ensure good aquarium plant growth without encouraging the excessive growth of unwanted, unattractive green algae – usually a sign of too much light.

Warning

When dealing with electrical equipment, particularly near water as in the aquarium, make all connections and adjustments with the supply disconnected from the mains.

Filtration

In nature, fishes generally swim away from trouble (if they are quick enough) and almost certainly will select the best part of the river, stream or pool in which to live, where the supply of food is greatest or the water condition is cleaner and healthier. In the aquarium, the fishkeeper has to bear the responsibility for both the supply of food and the cleanliness of the water. The latter is taken care of by efficient filtration, often coupled with regular partial water changes.

The need for filtration
Water in the aquarium becomes contaminated by the fishes' waste products; in addition to the physical wastes there are also chemical actions occurring which must be kept under control to avoid the inevitable build-up of toxic materials. The most dangerous of these are ammonia and nitrite (an ammonium compound), both of which are toxic to fishes in varying degrees. Carbon dioxide is also dangerous if allowed to build up, but we shall see that this is easily dispersed with aeration and by the movements of water also brought about by the use of filters. (The action of aquatic plants in the dispersal of carbon dioxide is described later. See the section on plants on page 26.)

Filtration of aquarium water can be achieved by mechanical, chemical or biological methods.

Mechanical/chemical filters
Filters draw water through a container in which a filter medium (such as acrylic fibre) traps suspended sediment in the water. Activated carbon and zeolite may also be used to remove dissolved waste products and ammonia. The cleaned water is then returned to the aquarium by means of an airlift or an electrically driven water pump. Modern filters often incorporate especially designed filter medium 'cartridges' for easy removal. Although biological filtration works on a different principle, using suitable filter media can allow it to occur within mechanical systems.

Mechanical filters can be of

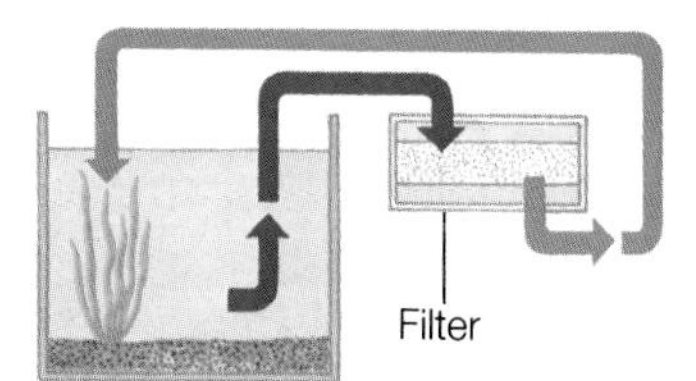

Above: *The principle of mechanical filtration. Dirty water is removed, cleaned and returned to the tank.*

Left: *This canister-type power filter can be situated away from the tank. Note the isolating hose taps for easy cleaning.*

Left: *The electric impeller fitted to this filter provides a much faster water flow than in air-operated types. The open-box design makes cartridge replacement easy.*

several designs. For small aquariums, an inexpensive internal bottom filter powered by air from an air pump is frequently used. As an alternative to the operation by air, motor filters use a small electric water pump to achieve circulation. These filters may take the form of a box hanging on the outside of the tank, a submerged filter, or a sealed canister.

Motor filters can give a relatively small flow but larger models are ideal for large tanks or for those that contain fishes with known dirty habits. One virtue of canister motor filters is that they can be mounted remotely from the aquarium (often in a nearby cupboard or in the plinth on which the aquarium stands) and this helps enormously when trying to fit an aquarium into a room's decor as artistically as possible.

Biological filters

Biological filtration converts the toxic waste products produced by the fishes into relatively harmless compounds. This is achieved by cultivating bacteria, either in special filtration media for mechanical filters or on the gravel in undergravel filters, which are described below.

Nitrosomonas and Nitrobacter bacteria respectively change ammonia to nitrite and then nitrite to nitrate; the latter is virtually non-toxic to fishes and can also be utilized by plants. Modern semi-dry trickle filters can convert the remaining nitrate back to free atmospheric nitrogen, so completely removing waste material of this nature from the water.

Undergravel filters consist of a perforated plate to which is attached an air tube. This system causes water to flow through the gravel, which is placed on top of the plate. The airlift is now frequently replaced by a small electrically powered pump known as a powerhead.

Below: *The principle of undergravel filtration. Water passes through the gravel (placed on top of the plate), where a colony of bacteria dentrify the waste products.*

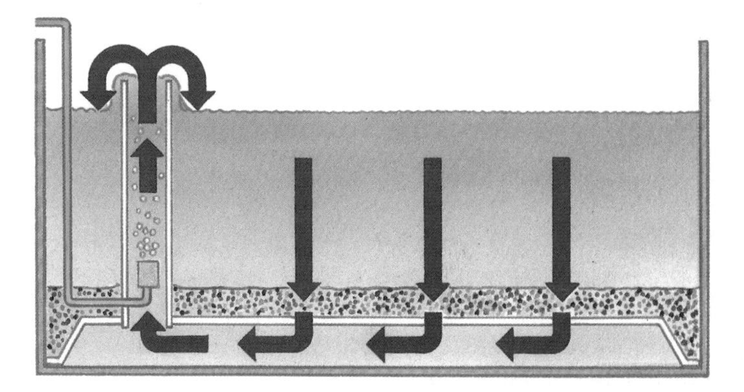

Above: *As its name suggests, the undergravel (biological) filter should be installed before the gravel. The filter plate should fit the tank as exactly as possible to provide the maximum filtration area.*

Below: *The depth of gravel (particle size 2-3mm/0.08-0.1in) should be at least 5-7.5cm (2-3in). This allows for optimum water flow through the gravel and good plant growth. Slope the gravel from back to front.*

Depending on the design, water is pumped either down or up through the gravel and this flow of oxygen-carrying water maintains on the gravel the colony of bacteria that clean the water.

Undergravel filtration is safe to use, even with newly born fishes, and being beneath the gravel is not unsightly. Heeding the following points will ensure the filter's overall efficiency.

The depth of gravel above the filter plate must be at least 5-7.5cm (2-3in) and ideally the filter should cover the whole of the aquarium base. There should be no way for water to short-circuit the system, eg around the edges of the plate, and some fishkeepers advocate sealing the edges of the filter plate when installing. Fishes that dig in the gravel will, in time, uncover the filter plate, thus nullifying the filter's action. This can be guarded against by placing a piece of nylon netting on the first 2.5cm (1in) of gravel before the final top layers are added.

It is possible to build undergravel filters external to the aquarium, and enterprising fishkeepers will also realize that a filtration system incorporating both mechanical and biological filtration is fairly easy to design and build, although its final appearance may not be appreciated by all who see it.

Using air in the aquarium

Air is supplied by the air pump and is used to create turbulence and water circulation by means of an airstone, or it can simultaneously be used to operate filtration systems.

Modern air pumps are quiet and reliable (quietness usually increases with the price) and may be of two types, vibrator or piston operated. Vibrator pumps are inexpensive and

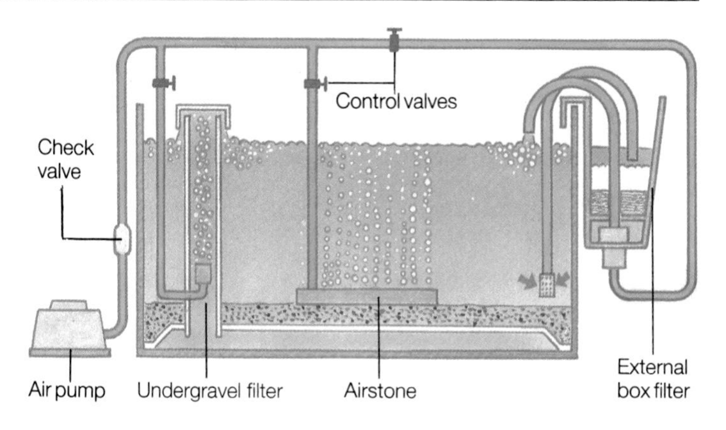

Below: *A set of ganged (multiple) air valves can be fitted to the side of the tank for easy access. The amount of air reaching each piece of equipment in the aquarium can be controlled by setting up every valve separately.*

Above: *The output from the air pump can be distributed to the various filters and airstones by plastic tubing and regulated independently by control valves. The check valve guards against back-siphoning of water.*

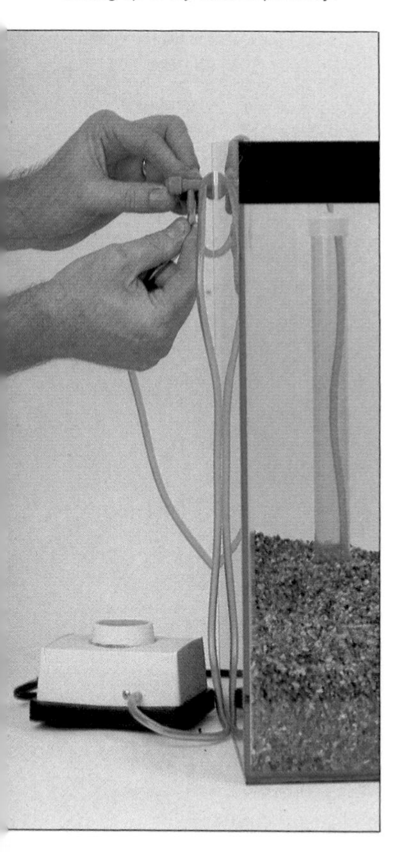

are quite adequate for a community tank owner. Piston pumps can be more expensive than vibrator types but they run with a more acceptable noise to some fishkeepers' ears.

All pumps require periodic maintenance; it may be the cleaning of the air filter, the replacement of the rubber diaphragm in the vibrator pump, or lubrication of the bearings and cleaning the oil filters in the piston pumps. Both types of pump need to be safeguarded against water siphoning back into them if the electricity supply is interrupted. Pumps should be sited above the water level of the aquarium;failing this, the airline from the pump should be raised a few centimetres above the water level in the tank before being connected to the airstone, filter or air control valves.

The rate of air flow required is usually controlled by control valves, clamping screws or a control fitted directly to the more expensive vibrator pumps and incorporated in their design. Excesses of air from the air pumps should be bled off to the atmosphere and not constricted by clamping the airline, as the resulting back pressure created may well cause excessive mechanical wear to the pump unit.

Tank furnishings

For the purposes of studying a fish's anatomy closely a small bare tank is ideal, but in order to study a fish's way of life with all its complexities of feeding, territorial possessiveness, breeding etc it is necessary for the fish to feel as at home in the aquarium as it would in its natural surroundings. Hence, the tank needs to be furnished with gravel, rocks and plants to provide the fish with shelter and a sense of security, for illogically the more retreats you offer a fish the less they seem to be used.

Gravel and rocks

One of the most reassuring environments a fish can have is a dark aquarium floor-covering; seen from above, the fish's dark top surface merges excellently with a dark background, whereas if light-coloured sand is used every fish is conspicuous. The usual gravel obtainable from aquarium dealers, of a dark to mid-brown/yellow appearance, is quite suitable for a community collection. The particle size of 2-3mm (0.08-0.1in) is recommended, being neither too coarse (which would allow food to be trapped uneaten within its interstices) nor too fine (which would hamper plant root growth and also impede water flow through a biological filtration if this is fitted).

Many fishes, particularly nocturnal species, enjoy resting places, and rock caves provide ideal sanctuaries. In addition, rocks help to break up the uniformity of the aquarium, and by careful use of rocks different levels of gravel banks can be created and maintained. Only non-soluble, hard rocks with no evidence of metal ore-bearing veins should be used; limestones and calcareous rocks are not suitable. Slate, once any jagged edges have been removed, can be structured into layered outcrops (a dab of silicone adhesive will keep the desired shape in place), and if cave-

Below: *There is a wide range of natural and synthetic materials available with which to furnish the aquarium, some decorative and some functional for the fishes' needs.*

Right: *This furnished aquarium combines gravel, slate, synthetic logs and plastic plants to produce a striking and stable environment for its mixed population of fishes.*

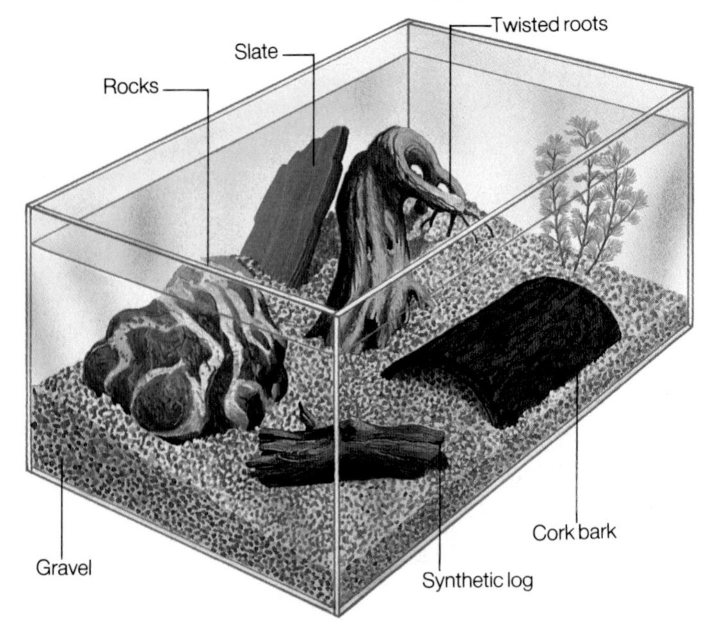

or rock-dwelling fishes are to be kept, this is an easy way to make sure that each fish has its own territory. To disguise the aquarium's boxiness the sides and rear walls can be covered with pieces of rock, again stuck on with silicone sealant. Another way to make the aquarium seem bigger is to build a rocky diorama which is fixed (and lit) behind the aquarium itself.

Other tank furnishings

Other fishes naturally inhabit reed-beds and areas where submerged tree roots offer refuge. Such conditions can be created within an aquarium by the use of wood and dead roots, preferably collected from a river bank where the waterlogging process has rendered them suitable for aquarium use. Any wood used in the aquarium must be free of decay and boiled in several changes of water before use; alternatively, wood can be given several coats of polyurethane varnish to seal it and prevent tannins from leaching out into the aquarium water.

Cork bark is another favourite aquarium decoration and can be shaped and cut easily. Natural-looking logs can be made from rolled cork, and can be persuaded not to float up by the use of a gravel-filled pipe or a base plate buried in the gravel on the aquarium floor.

Synthetic aquarium decorations may be easier to acquire than real logs, and these artificial but realistic-looking substitutes soon take on the appearance of the real thing as they become covered with living algae.

Decorating the aquarium is a compromise between the needs of the fishes and what satisfies your artistic needs; remember, you are trying to create a natural environment for the fishes rather than a work of art for yourself. However, if you put the fishes' requirements first, you will be even more pleased with the finished effect of your furnished aquarium.

Water

To us, water is just another commodity and we are quite unconcerned with its quality, taking for granted that as long as it comes out of the tap in sufficient quantities for our wide-ranging uses it is also safe and unpolluted.

To a fish, water is everything and any deterioration in its quality will immediately affect the fish's whole lifestyle and, perhaps, its very existence.

Partial water changes

We have already considered what effects keeping living animals in the confined space of an aquarium has on the water's cleanliness and purity, and how this can be maintained through the use of a filtration system.

Another simple method of keeping sediment and dissolved wastes down to a low level is by carrying out partial water changes; the replacement of dirty water with fresh clean water dilutes the remaining aquarium water. Approximately 20-25% of the aquarium water should be removed every 3-4 weeks, using a siphon tube; water should be drawn from just above the surface of the gravel, thus removing settled sediment and detritus at the same time.

The replacement water should, ideally, be of the same temperature as the aquarium water, to avoid chilling the fishes unduly; many fishkeepers add fresh water straight from the tap and say that their fishes appear to enjoy the cooling experience, swimming in and out of the water stream quite happily. It is probably true that fishes can withstand such shocks to a certain degree. In nature they may well experience such temperature changes as, for instance, melting snows deposit cooler waters into streams and rivers each spring. The problem is that although fishes from mountain streams are hardy enough to stand such treatment, other species in a community collection, from much warmer climes, may be stressed at water change times. When a fish becomes stressed (for whatever reason) it risks disease, and once it contracts an ailment this soon spreads throughout the aquarium. It pays to think a little, before taking any short-cuts that might endanger your fishes' continuing health.

Using tap water

Can we keep fishes in water from the

Below: *Many of the most popular tropical aquarium fishes come from this environment – the soft waters of the Amazon Rain Forest region.*

domestic supply? Yes, we can, fortunately, otherwise the aquarium hobby would be in a very sorry state if we had to collect water (as well as fishes) from the wild! However, there are one or two precautions to be taken into consideration.

Water from the domestic supply is intended primarily for human consumption and has been heavily treated for this purpose. Also, the water is stored for some time within the distribution pipework both outside and inside our homes. Therefore, in addition to treatments added deliberately, water also picks up metals from the pipes in which it is conveyed and stored. When water is drawn from the supply for use in an aquarium, the first few litres of water should be run to waste; once this water that was standing in the pipeworks has been discarded, the 'newer' water can be used to fill the tank. The water in the tank should then be vigorously aerated for several hours to drive out the chlorine in the water. Alternatively, dechlorinating agents can render the water safe for immediate use, and will also precipitate heavy metals such as copper and zinc from the water.

In a community collection, it is impossible to give each species the exact water conditions that it enjoyed in nature, but the fishkeeper ought to familiarize himself with some basic water chemistry in order to understand the needs of the fishes, especially in later years if attempts are to be made to breed those species to which water conditions are more critical for full health.

Although pure in its original form as rain, water is soon polluted by gases in the atmosphere and by minerals in the ground. These pollutants give water two main characteristics, which can be measured in order to compare one water's quality against another's; these characteristics are pH (the degree of acidity or alkalinity) and hardness (hard or soft water).

The pH of water

Although the pH scale (measuring acidity or alkalinity) ranges from 0 to 14 (strongest acid to strongest alkali, with pH7 as the neutral point), the range that concerns freshwater fishkeepers is limited to around 6.5-7.5, the range of pH conditions approximately covering the waters in which our fishes are found. (The pH of the domestic water supply is between

Below: *A view across Lake Malawi, one of the Rift Valley lakes in Africa that supports a unique population of hard water Cichlids.*

Above: A reagent is added until a colour change occurs. The sample is then placed in the holder for comparison.

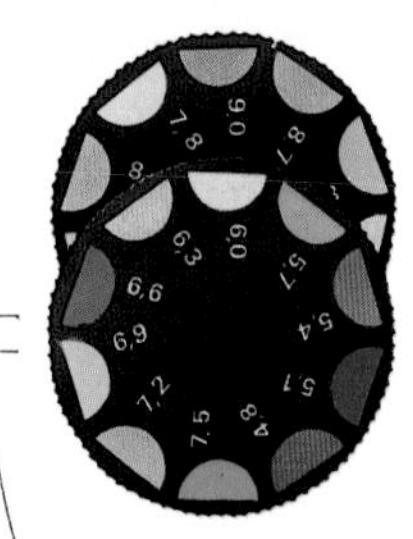

Above: The colour wheels are coded to show different pH values and are calibrated for either fresh or sea water.

6.5 and 8.5 depending on locality.)

The pH of water can be ascertained by the use of inexpensive test kits, which work on a colour comparison basis. For the community aquarium there is no need to delve into the pH of the water to any degree, if at all; pH is not stable, but changes slightly throughout the day. For this reason, pH readings should always be taken at the same time of day and at the same water temperature. The alteration to the water conditions, to give a precise pH requirement, is something that a community collection will not need, and is too complex a subject to be included in the Guide. Many fishkeepers confuse pH with hardness; the two are only very loosely related as far as aquarium matters are concerned. Generally (which means there are exceptions!) acidic water (below pH7) is soft, and alkaline water (above pH7) is hard.

Water hardness

Hardness is due to dissolved salts, usually of calcium and magnesium, and can be either temporary or permanent. Temporary hardness can be removed by boiling but permanent hardness can be removed only by distillation or by using ion exchange methods. A simpler way of reducing both forms of hardness is by dilution; the addition of pure water (no hardness) or of a less hard water will

Above: *Test kits for measuring the pH of water are easy to use and cover the pH ranges likely to be found in the average tropical aquarium.*

produce an overall softening of the water. (Partial water changes will not soften water if the replacement water is the same hardness as the water removed from the aquarium.)

Although fishes come from waters of different hardnesses in nature, they are generally able to acclimatize themselves to a common set of water conditions without too much trouble. Most fishes caught in the wild are subjected to several changes of water conditions before they reach our aquariums, and their final stay (at your local aquarium dealer) will finally accustom them to your local water conditions.

The aquarium water's hardness figure can be checked using a test kit, in a similar way to testing for pH. Hardness may not fluctuate quite so rapidly as pH, but it is very dependent on what minerals are present in the aquarium's furnishings. A high calcium content in the gravel or rocks will harden the water over a period of time, and a close check must be made on anything put into the aquarium if carefully prepared water conditions are not to be upset. There are other confusing, and often obsolete, units of hardness

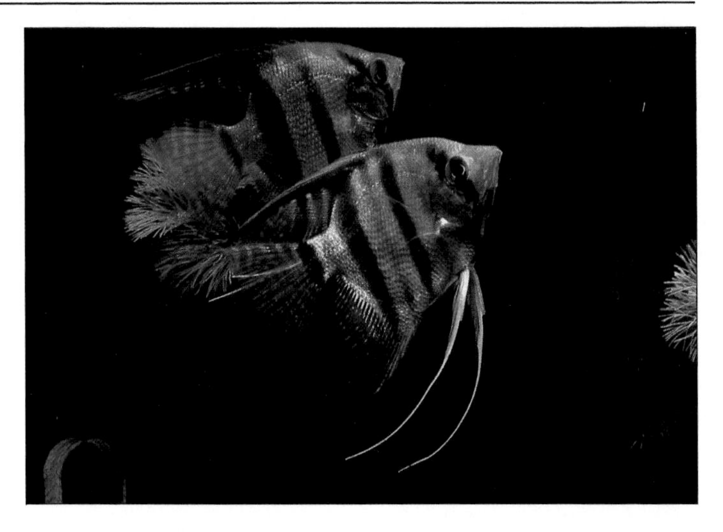

Above: *These attractive Angelfish* (Pterophyllum scalare) *are native to the River Amazon in South America and prefer soft water in the aquarium.*

Below: *This very colourful Platy* (Xiphophorus variatus) *lives naturally in the rivers of Mexico and thrives particularly well in hard water.*

measurement; some methods base their calculations upon amounts of calcium carbonate (parts per million) in the water, others measure the calcium oxide (milligrams per litre). Most European hardness test kits measure the calcium oxide in the total Carbonate Hardness figure, and the non-Carbonate Hardness figure is found by subtracting the Carbonate Hardness figure from the General Hardness figure. (Carbonate and General Hardnesses are determined by separate test kits.) As a rough guide to the scale of hardness 0°-3° is very soft, 4°-7° soft, 7°-12° hard, and 13° upwards very hard.

To sum up, attention to water cleanliness and temperature is very important to all fishkeepers, no matter what species they are keeping; to the community collection, scientific matters such as pH and hardness are not so vital but may be investigated should the fishkeeper's curiosity require it. Knowledge of any aquarium subject is never wasted, but the newcomer need not worry that such technicalities are obligatory from the start. Many fishkeepers have continued success because they do the right things without ever knowing it, while all the experts somehow contrive to make really hard work of it!

Plants and layouts

It is surprising that although the fishkeeper is usually knowledgeable about the fishes in his collection, the aquatic plants (which are so important to the aquarium) are more often than not considered as only part of the aquarium's furnishings. It is often forgotten that plants for the tropical aquarium come from the same exotic localities as do the fish, and are every bit as interesting.

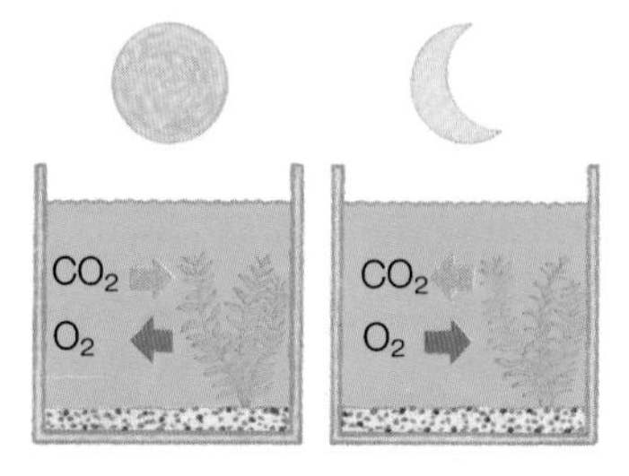

Above: *During 'tanklight' (left) all the aquarium plants absorb carbon dioxide and release oxygen; in darkness (right) the reverse occurs.*

Below: *Artistically minded aquarists may aspire to aquascaping their aquariums to this high standard of colour, shape and composition.*

The value of aquatic plants
The various shades of green (and sometimes of red), together with the many different leaf forms and sizes of the aquarium plants can be set out in the most stunning arrangements, often more densely planted than nature intended. To the fishkeeper, and to any appreciative onlooker, the visual impact is the major attraction. To the fishes, however, the aquarium plants provide much more important and fundamental services.

Plants provide shelter and territorial boundaries. Floating species, or the leaves of long plants trailing over the water surface, bring welcome shade and a refuge for young fishes. At breeding times plants may be used as spawning sites or as basic building material for bubble-nesting fishes. Vegetarian species, if not provided with enough green food by the fishkeeper, will turn their attention to any soft-leaved plants growing in the aquarium.

In addition to these, there is yet another service that plants render which is directly contributable to the continuing health and success of the aquarium, and that is the control of carbon dioxide. During daylight – or, in the case of aquariums, the hours in which the tank lights are on – plants build up their internal food resources within their green chlorophyll cells by photosynthesis. During this process carbon dioxide is extracted from the water, and oxygen released. Also, nitrates formed at the end of the denitrifying action by bacteria (see Biological Filtration, page 17) may be taken up by the plants. Many plants absorb salts through their leaves; their roots are used only to anchor them in the aquarium gravel.

During the hours of darkness, the plant uses up its food stocks, photosynthesis does not occur, and the plant respires in the same way as the fishes, ie oxygen is consumed and carbon dioxide is expelled. In densely planted aquariums with large numbers of fishes there may be large amounts of carbon dioxide produced at night, and aeration should be provided at the water surface to assist dispersal of the carbon dioxide.

Selecting plants for the aquarium is very much a personal choice but each type of plant should be chosen for a specific purpose or part of the aquarium layout. Plants can be conveniently divided into three groups: rooted species, floating species and cuttings.

Rooted plants

By far the largest number of species come from the first group, within which are almost all the shapes and sizes one could wish for. A sensible guide is to use large numbers of plants from a small number of species; plants look more natural in clumps of similar species rather than a haphazard mixture of different plants. Similarly, planting arrangements should follow commonsense rules, with low, slow-growing plants covering the gravel at the front and bushy and taller species filling out the corners, back and sides of the aquarium.

Not all plants require the same amount of light, and shorter plants may usefully be planted in the shade

Above: Vallisneria, *an extremely useful background plant for the aquarium. Justifiably popular.*

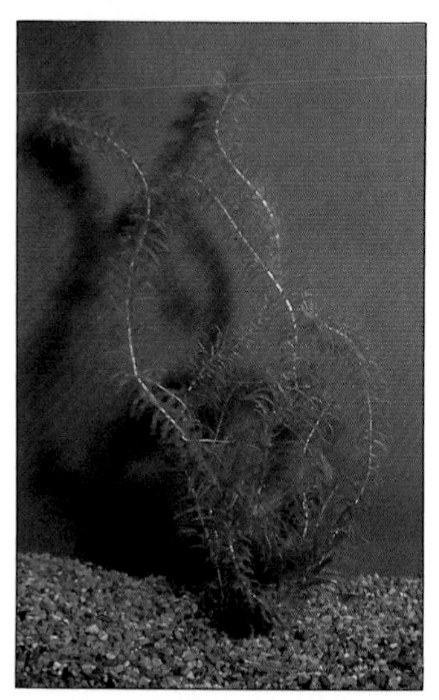

Above: Egeria densa, *a vigorous aquatic plant that quickly forms a dense mass. A superb oxygenator.*

of their taller neighbours. To add dramatic interest, the aquarium should feature a few specimen plants, which are planted not in clumps but in isolation, in front of rocks etc. Such stars of the aquatic plant world may be grown in individual miniature pots, which are buried in the gravel; in this way, the plants can be given special treatment such as tablet plant foods, or planted in special compost, and will not be disturbed either by the digging of fishes or by being transplanted (complete with pot) to other aquariums by the fishkeeper.

Floating plants

Floating plants are usually fast growers and may require thinning out from time to time. Most have relatively long roots that hang down into the water, and these provide an excellent refuge for very young fishes. The floating leaves should be protected against scorching by the aquarium lights (particularly filament-type lamps) by the use of a cover glass fitted on top of the aquarium below the reflector/hood. This cover glass will also protect lamps from splashes and prevent fishes from leaping out.

Cuttings

Plants grouped under this heading are usually those fine-leaved species that form bushy growth. Although they have roots, these develop from points on the stem, and hence any cutting taken from the plant will soon root of its own accord when replanted in the gravel. This is a convenient way not only of increasing the numbers of these species but also a way of encouraging the donor plants to develop into luxuriant bushes, as the pruning and taking of cuttings encourages side-shoots to grow. Some broader-leaved species also root readily from cuttings, and even a single severed leaf may develop roots if left floating in the aquarium for a few days. It can then be planted.

How aquarium plants reproduce

Because of their permanently submerged life, most aquarium plants reproduce vegetatively. They send out runners from the parent plant and small new plants develop either at intervals along the runner or at the end of each runner; when the new plants have grown two or three leaves they can be disconnected

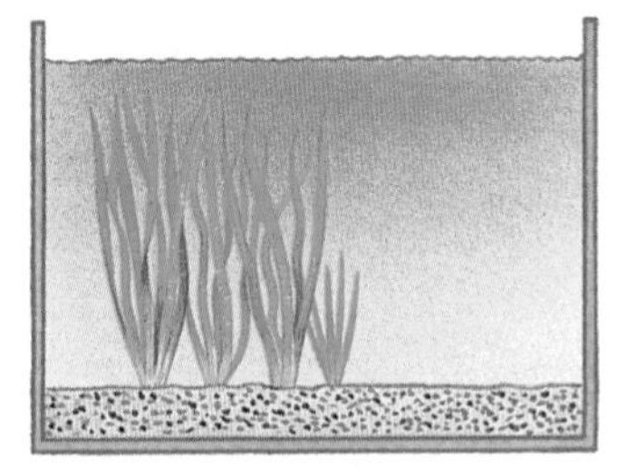

Background fillers
Sagittaria, Vallisneria. Tough leaves. Propagate by runners. These plants sometimes produce flowers.
Types of fish: *Angelfish, Gouramies, Cichlids, Headstanders, Killifish.*
Spawning value: *Negligible as sites, but these plants will provide a refuge for the female to escape the aggressive attentions of the male.*

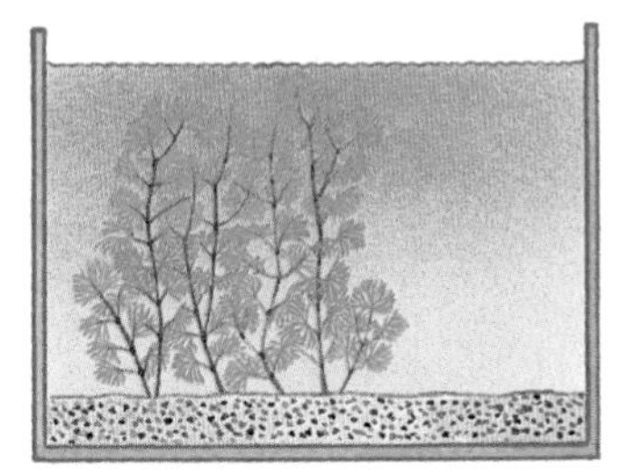

Space and corner fillers
Ambulia, Cabomba, Ceratophyllum, Egeria, Vesicularia. Soft leaves; may be browsed upon by some fishes. Can be propagated by cuttings.
Types of fish: *Barbs, Characins and Rasboras (except vegetarian ones).*
Spawning value: *Excellent aquarium plants for egg-scattering species and for some of the Killifishes.*

Specimen plants
Aponogeton, larger Cryptocorynes, Echinodorus. Tough leaves. Propagate by runners, division or, in some cases, from setting seed.
Types of fish: *All species*
Spawning value: *Useful aquarium plants for some Cichlids, Rasboras and other species that deposit eggs on or under plant leaves.*

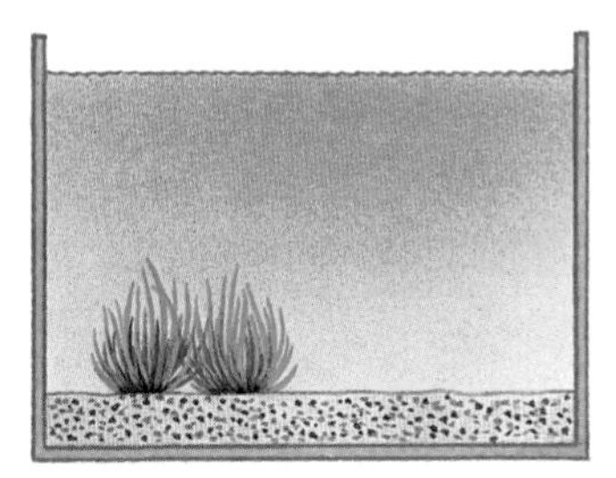

Low-level plants
Acorus, smaller Cryptocorynes, Microsorium. Tough leaves. Propagate by runners or division.
Types of fish: *Botia, the dwarf Cichlids, Corydoras.*
Spawning value: *Although these types of plants are of little value as spawning sites they may provide refuges for females or young fry.*

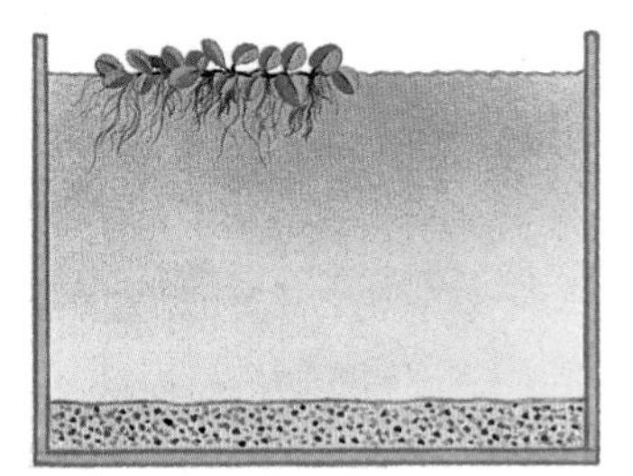

Floating plants
Azolla, Lemna, Pistia, Salvinia. Leaves floating above the surface; roots trailing down into the water.
Types of fish: *Danios, Gouramies, Killifishes, Live-bearers.*
Spawning value: *Often used in the construction of bubble-nests by Gouramies. Provide refuges for fry, especially newly born live-bearers.*

Above: Aponogeton crispus, *a fine species with ruffle-edged leaves. Note rhizome typical of this genus.*

Above: Acorus gramineus, *a hardy, slow-growing foreground plant best suited to the cooler aquarium.*

from the parent plant and rooted elsewhere. Other species will grow daughter plants directly on their leaf surfaces; these usually float free and can then be planted elsewhere.

Flowers on aquatic plants are not as rare as might be expected; very often a look below the reflector will reveal tiny blooms thrust above the water surface. In some species the fishkeeper can pollinate the flowers and obtain seeds that can be sown in shallow warm water to produce more plants to furnish new tanks.

Algae – a persistent problem

One plant that most fishkeepers can grow without any effort at all, and to their dismay, is algae. This pest may take on many forms, from a slimy blue-green covering to a tangled mass of what looks like green candy floss. Generally, these growths occur where there is an excessive amount of light, or where there are not enough

Right: *Although there are still some gaps visible between the plants in this newly furnished aquarium, these will soon close as growth continues in this well-managed environment.*

Above: Pistia stratiotes, *one of several floating plants available that provide convenient shelter for fry.*

aquarium plants to crowd out these aquarium weeds. There will always be a soft covering of algae on the walls and front glass of the aquarium; the front glass should be scraped clean, but algae (except blue-green) on the other surfaces should be left undisturbed for the benefit of those fishes that like to graze on greenstuffs – it might as well be the algae, in preference to the plants!

Planting the aquarium

Planting an aquarium is best done with the tank three-quarters full of water. As soon as they are planted the majority of plants then take up their natural growing positions and the effects of your aquascaping can be seen immediately; the view through the front glass is always foreshortened by the water and this gives a different perspective compared to a tank that has been planted dry. Species that have a conventional root system should be planted in such a way that the junction between leaf stem and root is just above the top of the gravel. Brunches of cuttings can be weighted down with lead strip until rooting occurs; the lower section of the stem should be stripped of leaves before burying cuttings in the gravel, and roots will develop from these previous leaf nodes. Cuttings can be protected against uprooting by fishes by placing a few pebbles around their stems to hold them in place.

Unwelcome visitors

Apart from vegetarian fishes, another threat to plants comes from snails, which may devour many of your plants or leave them looking tattered and torn; although there are many 'cures' for snails, ranging from snail-killing additives to electrocution with torch batteries, the only real defence is not to allow them access to your aquarium in the first place. All plants for the aquarium should be checked for snails and snails' eggs, before being used. Some fishkeepers give plants a rinse in a weak solution of potassium permanganate to disinfect them and to kill any other unwanted visitors that may be clinging to them.

Feeding

Thanks to years of research and development, fishes in captivity are probably better fed than their relatives in the wild, and do not have to expend half as much energy in getting their food. Aquarium fishes receive their food by courtesy of the fishkeeper, and the responsibility for the quantity and quality of the food is entirely his. A fully varied diet is as necessary for, and appreciated by, the fishes as it would be for us, and such a diet ensures that they receive all the proteins and vitamins they require, which might be lacking in a more restricted diet.

In understanding fishes' dietary needs, we should also take note of how each species takes its food and from what position in the water. Surface swimmers use their upturned mouths almost like a scoop, their forward movement forcing food into their mouths as they patrol just beneath the surface. Fishes with terminal mouths may well snatch their food from the surface in isolated dashes but generally food is taken (often greedily) as it sinks through the water. An underslung mouth is best for foraging around the aquarium floor or for browsing on algae-covered rocks or aquarium glass.

With these feeding habits in mind (and there will be some of each within the community collection) it will be quickly appreciated that no one food will suit all fishes. A floating food will never reach the bottom-dwelling fishes, and a rapidly sinking food will be lost to surface swimmers. A compromise can be easily reached by using different types of food – flake, granular or tablet – spread between the different feeding times. It is no use giving one heavy feed of all types of food and hope the fishes will somehow sort it all out and even save some for later.

Now is the time to learn probably the most important basic rule in fishkeeping: DO NOT OVERFEED.

Manufactured foods

The majority of food given is manufactured, not living, although we shall be looking at live foods shortly. Manufactured foods have one drawback among their many excellent advantages; they

Below: *These vivid male Guppies are feeding from a freeze-dried Tubifex cube stuck to the aquarium glass. Use more cubes (stuck at any depth) for a larger population of fishes.*

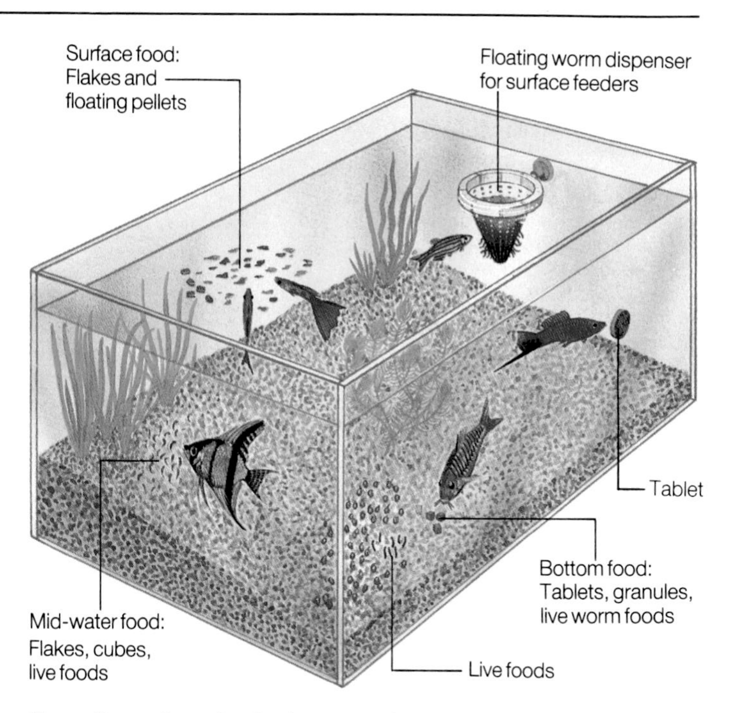

Above: *Types of aquarium foods. Flake foods float for a short time before sinking slowly. Pellets do not sink. Freeze-dried cubes can be stuck to the glass or kept in weighted plastic baskets for bottom feeders. Floating worm dispensers can be used to provide a supply of live worms (Tubifex) at the water surface for the top-swimming fishes.*

decompose in water if left uneaten. This decomposition leads to pollution of the aquarium water, which will eventually kill the fishes. Fishes should be given only as much food as they willingly eat in 3-4 minutes, and this amount is very small. It is better to give little and reasonably often, and a hungry fish is always healthily on the lookout for food.

Feeds are usually given at the convenient times of morning and early evening to coincide with the fishkeeper's visits to the aquarium before and after the working day. A little more food, last thing at night, of a suitably fast-falling variety, will provide an almost exclusive meal for nocturnal bottom-dwelling fishes. Incidentally, all fishes should be treated as equals. Many bottom dwellers are kept in the belief that they will tidy up the aquarium floor, and this they will do, but they should also be considered worthy of having food given especially for them.

Live foods

So far, we have discussed manufactured foods but obviously fishes in the wild are more used to taking living food – insects, waterborne crustaceans, larvae and smaller fishes. Aquarium fishes may also be given live foods and these fall into two groups, those that are naturally aquatic and those that are not, such as earthworms.

Aquatic-living foods include the water-flea, *Daphnia*, mosquito or gnat larvae, *Tubifex* worms and numerous other tiny inhabitants of pond or rainbutt. Any wild-caught live foods should be carefully checked for unwanted 'visitors' among their numbers before being put into the aquarium. Larvae of the dragonfly, water boatman, great diving beetles, leeches and, of course, snails should be looked for and removed. There is also a danger that living food caught from fish-populated waters may also carry fish diseases.

Non-aquatic live foods include a number of worms of the Enchytreidae family. These range in size from tiny hair's-breadth proportions up to stout thread size; known to fishkeepers as micro-worms, grindal worms and white worms (in ascending size order) they can be cultured in boxes and fed usually on cereal-based foods. As one culture nears the end of its useful life (your nose will tell you when!) a new culture can be started with a seeding of worms from the preceding culture.

Earthworms (small ones whole, or large ones chopped up!) are relished by fishes. Earthworms are best collected from beneath a damp sack laid in a corner of the garden, or from a compost heap. Do not use worms collected from any part of the garden which has been treated with weedkiller or other garden chemicals.

Cultures of wingless fruit flies (*Drosophila*) make good food and are snapped up as they float on the surface. Supplies of this food may sometimes be difficult to obtain, but usually a member of the local aquarist society will have a culture or know the whereabouts of one.

Some aquarium dealers sell live *Mysis* shrimps, which provide an occasional treat for your fishes.

Modern technology has also developed an alternative means of preserving fish food – freeze-drying. In this way, live foods can be collected on a large scale and freeze-dried, and thus their inherent nutritious content can be given to the aquarium fishes whenever convenient, and not be restricted to seasonal catches. Just what advantages are lost by the fact that the freeze-dried foods are dead has not been discovered, as the fishes cannot tell us whether or not they prefer the real (living) thing.

Another source of food is the kitchen. Fishes will take all manner of household scraps including raw lean meat, peas, lettuce, spinach, pieces of raw fish, cod's roe, potato, cheese etc. Introduce new foods gradually to your fishes; try them first on hungry fish. Avoid all fatty foods. Remove any uneaten food; chunky foods can be dangled in the aquarium on a piece of thread for easy removal.

Right: *Angelfish feeding at the water surface. These fishes are always on the lookout for food and will even take food from the aquarist's hand.*

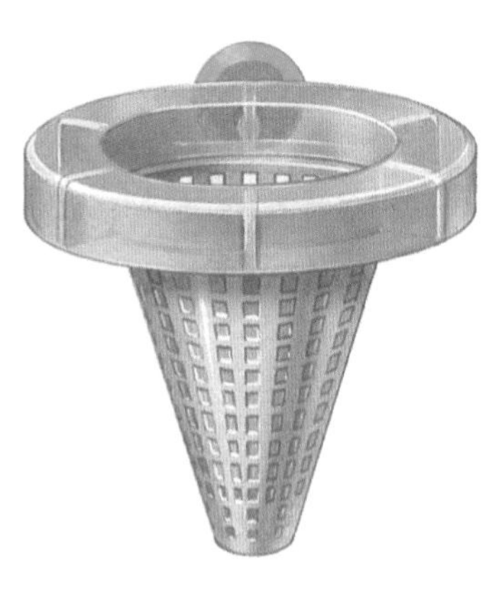

Above: *The worm dispenser is best fixed in position by a rubber sucker rather than allowed to drift around the surface of the aquarium.*

Above: *The top part of the worm dispenser can be used independently as a feeding ring to keep floating foods conveniently in one place.*

Foods for young fishes

Feeding newly born fishes used to be a messy and smelly business. Cultures of microscopic life (*Infusoria*) were encouraged by pouring boiling water over such things as chopped hay, crushed lettuce leaves, banana or potato skins etc. The resulting liquid was then supposed to burst into life with teeming millions of

Below: *Hatching brine shrimps. The dry eggs (1) should be added to a solution of sea salt and agitated by an airstone (2) until hatching occurs. The live shrimps (3) should be strained through a net (4) and further rinsed with fresh water before being used as a food. The salt water can then be used again to hatch more eggs.*

infusoria after a day or two's exposure to air and sunlight (a kitchen window ledge was a favourite advocated position). All that most fishkeepers got was the smell and a talking to by other members of the family! Nowadays, young fishes can be well catered for by both manufactured and live foods.

Manufactured foods are formulated in liquid, paste and powdered forms, for egg-laying or live-bearing young fishes. Food for young fishes should not be given until they are free-swimming (see Breeding pp 104-113) and can take the food, otherwise pollution occurs.

The best live food, especially for tiny fishes, is the newly hatched nauplii of the brine shrimp *Artemia salina*. Eggs of the brine shrimp can be stored indefinitely in a dry state, and when immersed in a saltwater solution (30gm/litre or 5oz/gallon sea-salt crystals) will produce tiny baby shrimps, which can be fed to young fishes. Again, progressive cultures should be made of these, to ensure a continuous supply of this nutritious, disease-free first food. The nauplii can be raised to bigger sizes for feeding to larger fishes; the food used to do this is normally yeast-based and convenient to handle.

Maintenance and equipment

Although the aquarium is well equipped with filters and generally self-functioning, a certain amount of periodic maintenance is necessary. Fortunately, most of this is an enjoyable routine and need not be looked upon with any reluctance.

Routine maintenance

Routine checks upon fish population numbers, water temperature and symptoms of disease all become automatic a few weeks after the installation of the aquarium. The growth (or otherwise) of the plants is also easily seen, and the chore of removing dead leaves is more than balanced by the taking of new cuttings and replanting fresh additions to the aquascape.

Filters require regular attention. The filter medium should be renewed every two to three weeks; some fibres can be washed through and re-used. There is no hard-and-fast rule when to change the filter medium, as its dirtiness depends upon so many variables: the size of the aquarium, the size and water flow of the filter

Below: *This reminder chart of regular jobs will help to ease those 'early days' worries. You may like to adapt it to suit your own particular needs.*

MAINTENANCE CHECK	Daily	Weekly	Monthly	Periodically
Check water temperature and number of fishes	●	●	●	●
Water condition Check pH				●
Partial change of water			●	
Filters Box filter: Clean and replace medium according to amount of use and state of aquarium				●
Undergravel filter: Rake the aquarium gravel gently				●
Plants Remove dead leaves and excess sediment on leaves; thin out floating plants		●	●	●
Prune; replant cuttings and runners as necessary		●	●	●
General Check air supply carefully; clean air pump valves and air pump filter				●
Clean cover glass				●
Remove algae from front glass of aquarium				●
Check fishes for symptoms of diseases				●

Note: If your fishes start to behave oddly, it may be worth checking over the tasks outlined above – regular aquarium maintenance can keep them healthy.

itself, how dirty the aquarium gets due to the actions of the fishes, and so on. Filter connecting hoses will become algae- and detritus-coated; a spiral brush mounted on a twisted wire handle is useful for cleaning these.

Electrically driven water pumps require regular cleaning and some models may require lubrication. The manufacturer's instructions will clarify this point. Similarly, the piston-type air pumps will require lubrication from time to time. Do not forget to clean the air filter in the base of the vibrator air pump if fitted. A reduction in air from the pump (usually accompanied by a clattering sound from the pump) is a sure sign of a split diaphragm. In the absence of excessive noise, the cause will be clogged air valves within the body of the pump. (Disconnect the electricity supply before dismantling the pump.)

The heating system is totally reliable, and works impeccably or not at all; there is no half measure stage. The thermostat, regardless of whether it is an internal or external fitting, has provision for adjustment of a few degrees either side of the factory set norm of around 24°C (75°F). Again, switch off before making any adjustments.

Cover glasses must be kept spotlessly clean so that the light is not prevented from reaching the plants. Experimentation may be made with different 'colours' of fluorescent lighting, different wattages of filament lighting or a combination of both types, to arrive at the most suitable lighting arrangement for your taste and for the best plant growth.

By far the most important regular task is changing a proportion of the aquarium water. This should be done every three or four weeks, and some 20-25% of the water must be removed. (See Water, p 22.) Sediment on the aquarium floor (also referred to as mulm or detritus), if not removed during partial water changes, can easily be removed by siphoning or by use of an air-operated sediment-remover. This is like a vacuum-cleaner which is swept across the gravel surface, the sediment being sucked up and deposited in a cloth bag. The clean water flows through the bag back into the aquarium.

The removal of algae from the front glass is quite simple. It can be scraped off using a razor-blade (some planting sticks have a holder for blades at one end), or by means of

Left: *This aquarium 'vacuum cleaner' removes sediment from the aquarium water easily and with little mess. Sediment is carried up in the flow of water generated by a stream of bubbles and is trapped in the cloth bag. The clean water passes through the mesh back into the tank.*

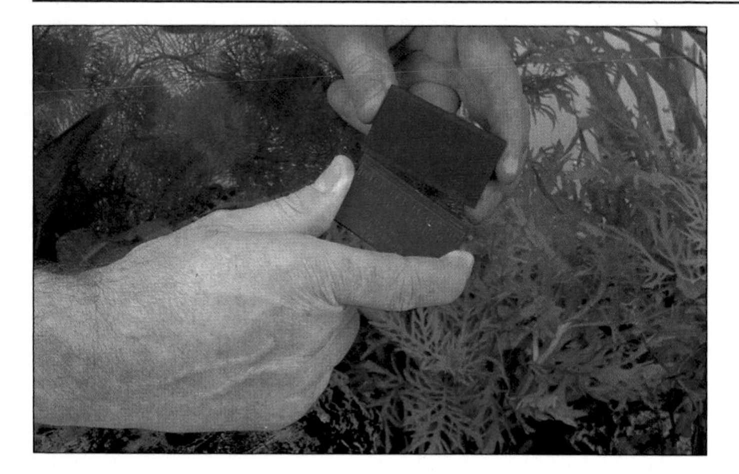

Above: *Magnetic algae scrapers, although convenient, should not be parked near the thermostat as they will affect the temperature control.*

Right: *Some fishkeepers prefer to use less sophisticated algae scrapers, such as the long-handled abrasive pad shown in use here.*

a magnetic abrasive pad, which is magnetically parked opposite its 'outside' handle when not in use.

Basic aquarium equipment
The most used piece of equipment is, of course, the net. After he has gained a little experience, the fishkeeper soon realizes the merits of some net designs over others. Rectangular or square nets with cranked handles make fish-catching much easier than round nets with straight handles. Some fishes are better captured with plastic bags, as their fins or barbels do not become enmeshed in the net. Nets made of soft silk are gentler on young fishes, and are also useful when fishing for live foods in ponds. Two nets are often better than one, as a fish can be guided into a large stationary net with a smaller one more easily than chasing a fish all over a planted tank with a single net, which does not do your patience or the plants in the tank any good at all!

Feeding rings and worm feeders ensure that food does not float all over the water surface, and that surface feeders get a chance to eat the worms before they would otherwise sink quickly to the aquarium floor. Both devices also train the fishes to expect food at the same place in the aquarium, and this is a useful opportunity to count and check the fishes as they congregate for food.

Opinion over the usefulness of planting sticks is divided, some fishkeepers finding their own fingers just as effective. This may be true during the initial planting of the tank, but later, when the plants have become fully established, there is no doubt that planting sticks are extremely helpful in getting into awkward corners to tend to the plants and insert new ones.

Air valves may be ganged together, which is more convenient than hunting along each airline for its controlling clamp. There is always one more air valve than appliances, the last one in the line being blocked off by a piece of airline tube (connecting both outlets together). This last valve can be used to bleed off excess air from piston-type air pumps as necessary.

Thermometers come in various forms: internal floating or with sucker, and external liquid crystal type. Whichever is used, it should be mounted so that the temperature is easily read, although most fishkeepers soon become adept at judging the temperature by the palm of the hand against the glass.

Below: *This mixed array of hardware includes some of the equipment necessary for setting up a tropical freshwater aquarium in the home. For newcomers to fishkeeping it may appear rather complicated but careful reading of books such as this will make it all seem clear.*

Selecting and introducing fishes

The selection of actual species must remain dependent on personal choice and local availability and therefore no directives as such will be given here.

What is more important is how to choose healthy stock and how to introduce them into the aquarium with the least possible fuss. Only by starting off with healthy, stress-free fishes can you give yourself and your fishes a better than even chance of making fishkeeping the success that it can be for anyone who tries it.

Selecting fishes

There are several outward signs to look for when buying fishes. The fins of freshwater fishes should be held erect and well away from the body; the body of the fish should be filled out, with no knife-edge profiles, and the backbone must not be bent or deformed. Fishes with wounds, pimples, boils, split fins, or clouded eyes are not a good buy, and any fish that obviously has difficulty swimming or maintaining a stationary position is also a poor bet. Certain species may not share their fellow tankmates' extroversion in parading themselves before the customers; these are usually the bottom-dwelling fishes, and often a closer inspection of the display tank (particularly around the base of the plants) will reveal interesting fishes otherwise ignored.

When selecting fishes for the community aquarium, you must remember that all fishes offered for sale are juveniles and will increase their size over the next few months. Unfamiliarity with the final sizes attained by some species will lead to problems later on, when that attractive little fish has grown up into an ugly big one with an appetite to match! The majority of fishes described in this Guide are suitable tankmates, although with one or two species tempers may become frayed at breeding time.

Another consideration to be taken into account is any one fish's sociability within its own species group. Some species are naturally gregarious, and thrive and look better in a shoal of, say, six fishes. A solitary

Above: *Whatever fishes you select from the bewildering display at your local dealer, always choose wisely – buy healthy, sociable stock.*

specimen often gains a reputation for harassing other fishes, but this may be the result of boredom and loneliness rather than any innate malice of the species in general.

At the outset, do not be tempted by very exotic (and expensive) species. These may well require special water conditions, food and extra care beyond the experience of the beginner. For the sake of the fishes, make haste slowly. If you have any doubts about the ultimate size of a fish or if you will be able to cope with its requirements, do not buy it; or alternatively, learn all about its needs from reading books or talking to the dealer first.

On the subject of dealers, cultivate a good relationship with one and you should be treated fairly. A responsible dealer will not mind being asked pertinent questions, as long as it is not done during his shop's busiest periods. A new arrival in the shop should not be snapped up immediately; a good dealer will often say, "Those have just come in, but leave them for a week or two to settle down." Also, ask what quarantine arrangements your dealer operates, so that you can be sure that any fish

Below: *Floating the fishes' carry home plastic bag in the aquarium restores their water temperature to the correct level before they are released.*

Below: *After 15-20 minutes the fish can be released. Give the existing inmates of the aquarium some food to take their minds off the newcomers.*

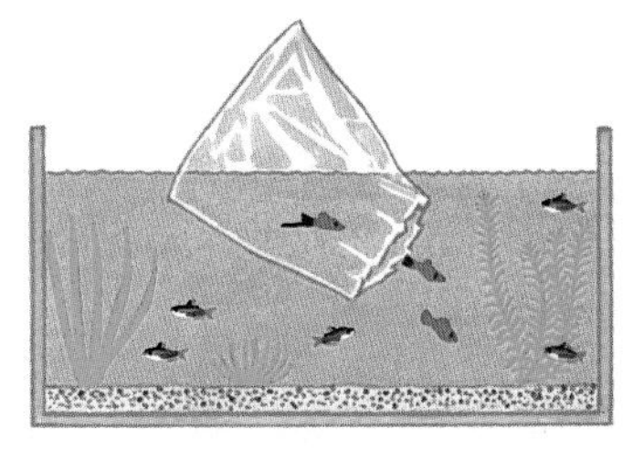

you buy is healthy and already acclimatized to the local water supply before you buy it.

Introducing fishes

Newly acquired fishes should be introduced into the aquarium in an unhurried way. The fish has had a fairly traumatic experience since leaving its country of origin, being passed from collector to exporter, exporter to airline, airline to importer, importer to dealer by the time it reaches your local shop. Fishes under stress are not so resistant to disease as others, and stress can be induced by careless handling.

Thanks to the plastic bag, fish transportation from shop to home is easy, but during the cold months of the year the bag should be well wrapped to prevent undue heat losses. Some heat loss is inevitable but the temperature of the fishes' water can be increased to that of the aquarium by floating the bag in the water for 20 minutes before releasing the fish. A specially designed floating acclimatization chamber is commercially available for this purpose also. To acclimatize the fishes to the *quality* of the aquarium water, small amounts of this can be added to the contents of the plastic bag during the temperature equalization period. In order that the new fishes are not harassed by the aquarium inmates some fishkeepers divert the aquarium fishes' attention away from the newcomers by giving food immediately prior to releasing the new fishes; alternatively, the lights can be switched off to hide the introduction of the new fishes.

Fish anatomy

Not every fish conforms to the traditional torpedo shape, and body shape reflects individual living and feeding habits. Ultra-streamlined bodies indicate fast-swimming, open-water predators, whose large tail fins are often complemented at the other end by a large tooth-filled mouth. Laterally compressed fishes such as the Angelfish (*Pterophyllum* sp.) inhabit slower flowing, reed-filled waters; and vertically compressed specimens live on the river bed itself.

The position of the mouth often indicates in what level of the water the fish generally lives. An upturned mouth indicates a swimmer just below the surface, whose mouth is ideally structured for capturing insects floating on top of the water. These fishes usually have a straight, uncurved dorsal surface. Fishes whose mouths are located at the very tip of the head, on a horizontal line through the middle of the body, are mid-water feeders taking food as it falls through the water, although they can feed equally well from the surface or from the river-bed, should the mood take them. Many other fishes have underslung mouths; and this, coupled with a flat ventral surface, clearly shows a bottom-dwelling species. But those fishes whose underslung mouths are used for rasping algae from rock surfaces (and the sides of the aquarium) may not be entirely bottom-dwelling. Some bottom-dwellers have whisker-like barbels around the mouth, which are often equipped with taste buds, so that the fish can more easily locate its food as it forages.

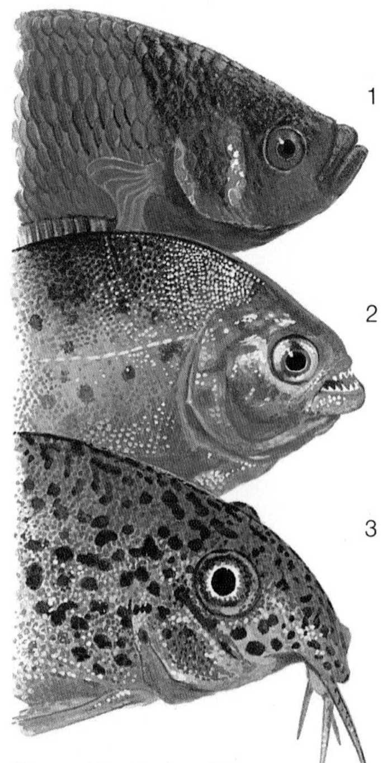

Above: *Mouths for all feeding purposes. (1) Upturned, as in top swimmers. (2) Terminal, typical of mid-water feeders. (3) Underslung, as in bottom-feeding fishes. The barbels have sensitive taste cells.*

The scales

A fish's scales provide not only protection for the body but also streamlining. A variation from a scale covering is found in the Armoured Catfish group (Callichthyidae), whose bodies are covered with two or three rows of overlapping bony scutes. Some catfishes, particularly the Mochokidae and the Pimelodidae, are 'naked' and covered in neither scales nor scutes.

The fins

The fish uses its fins for locomotion and stability, and in some cases as spawning aids either during courtship or in the hatching period of the eggs. Fins may be either single or paired. The caudal fin provides the final impetus to thrust the fish through the water, and fast swimmers have a deeply forked caudal fin. The Swordtail (*Xiphophorus* sp.) has an elongated lower edge to the caudal fin, in the male fish only.

The dorsal fin may be erectile (as in the Sailfin Mollies – *Poecilia velifera, P. latipinna*) and will often consist of hard and soft rays. In some species two dorsal fins may be present, but these should not be confused with the adipose fin, a small fin (usually of a fatty tissue) that is found in some species, notably the Characoid group, between the main dorsal fin and the caudal fin.

The anal fin is another single fin mounted under the body just forward of the caudal fin. Mostly used as a stabilizer, in the male live-bearing fishes it has become adapted to serve as a reproductive organ. In some Characoid fishes the anal fin of the male carries tiny hooks that help to hold the two fishes together during the spawning embrace.

The pelvic, or ventral, fins are paired and are carried forward of the anal fin. In many of the Anabantid fishes (Gouramies) these fins are filamentous and are often used to explore the fish's surroundings. The Angelfish also has narrow, elongated pelvic fins, but these are not so manoeuvrable, nor are they equipped with tasting cells. The Armoured Catfishes in the *Corydoras* genus use their pelvic fins to transport their eggs to the spawning site.

Pelvic fins in some species of Gobies are often fused together to form a suction cup that anchors the fish to the river bed and prevents it from being swept away by the action of the fast-moving water currents.

Below: *This Characin* (Moenkhausia pittieri) *has the usual seven fins plus an extra one. Can you spot it? Yes, it's the small adipose fin.*

Below: *This drawing shows the basic anatomy of a fish. It is based on the species above, one of those that possesses the 'extra' adipose fin.*

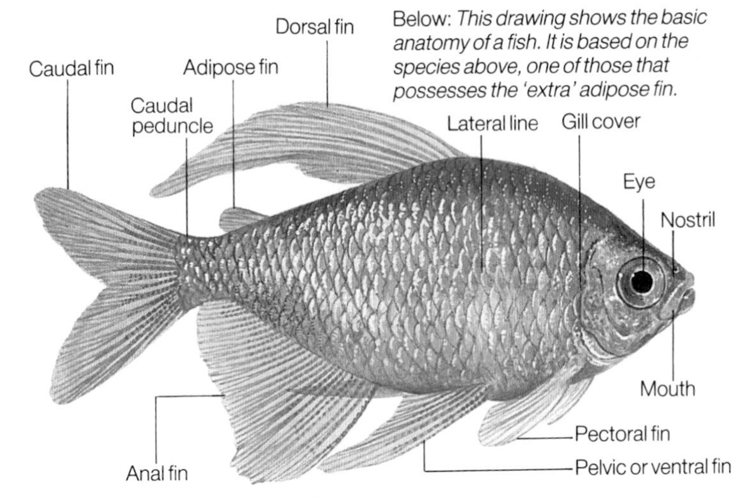

Above: *The Blind Cave Fish has no eyes (nor could use them in its natural underground cave waters) and navigates successfully using its pressure-sensitive lateral line.*

Pectoral fins emerge from just behind the gill cover (*operculum*). Primarily used for manoeuvring, pectoral fins have also been adapted for other uses. The Hatchetfishes emulate the marine Flying Fishes as they skim across the water surface by means of their well-developed pectoral fins. The marine Gurnard literally walks across the seabed on 'legs' formed by modified rays of its pectoral fins.

Many aquarium fishes have over-long, decorative fins. Fish breeders developed these exaggerated fins through deliberate breeding programmes and such fin developments are not found in their wild counterparts.

The fish's senses

The fish has the same five senses that a human being enjoys – sight, touch, taste, smell and hearing. Of these, the last two are more highly developed than those of humans. Many fishes detect food through smell, often over great distances. A fish's nostrils are not used for breathing, only for smelling. It is debatable where the sense of actual hearing ceases and the detection of low frequency vibrations begins in the fish world. This is because fishes are equipped with a sixth sense, the lateral line system. Through perforations in a row of scales, the fish's nervous system can detect minute vibrations in the surrounding water. This warns of other fishes or obstacles nearby. The Blind Cave Fish (*Astyanax mexicanus*) copes quite easily with life in an aquarium, navigating solely by means of its lateral line system.

Some fishes have developed sophisticated aids to help them cope in darkness or in murky waters, and these include the ability to generate a weak electromagnetic field. The Electric Catfish (*Malapterurus electricus* sp.), although scaleless, needs little protection against predators, because it packs a hefty electric shock. It is thought that it also uses this shock to stun smaller fishes.

The swim-bladder

A feature exclusive to fishes is a hydrostatic buoyancy organ known as a swim-bladder. This enables the fish to position itself at any level in the water, automatically giving the fish neutral density. Some fishes, notably the marine sharks, lack this organ.

Colour

Apart from attracting fishkeepers, colour plays an important role in the fish world. It serves to identify the species in general and the sexes in particular. It camouflages a fish from predators or gives clear visual warning that a species may be poisonous. Colour presents false targets to an attacker and gives some clue to a fish's disposition, showing that it may be frightened or angry.

Colour is determined by two

methods – by reflection of light, and by pigmentation. Those silvery, iridescent hues seen on the flanks of many freshwater species are due to reflective layers of guanin. Guanin is a waste product that is not excreted from the kidneys and body, but stored just beneath the skin. The colour seen depends upon the angle at which light hits and is reflected from these crystals. Many fishes, when lit by light coming through the front glass of the aquarium, seem to be coloured differently than when lit from directly overhead. This also explains why light-coloured gravel appears to wash out the fishes' colouring.

Fishes with deeper colours have pigment cells in their bodies, and some species are able to control the amount of colour they display. This can be seen quite easily in those species that tend to rest on the gravel surface or on rocks, where their colours are adapted to suit the background. Other fishes take on nocturnal colourations. The popular Pencilfishes (*Nannostomus* sp.) are notable examples and the hobbyist may be initially surprised at finding these fishes a different colour pattern each morning. Fishes effect such colour changes by contracting or expanding the pigmented cells (*chromatophores*) to intensify or dilute the colour showing through the skin.

Colour intensity is likely to be heightened in the male fish during the breeding period in order to attract a mate, and some female fishes within the Cichlid group may also have their colours exaggerated in order to be recognized by their subsequent offspring. A good example of this is seen in the *Pelvicachromis* genus, where the females are often more colourful during breeding than the male fishes.

It is possible to intensify fishes' colours by feeding them so-called 'colour foods'. These contain additives, such as carotin, that will accentuate colours. The Tiger Barb (*Barbus tetrazona*) is a favourite fish that responds quite startlingly to colour feeding, each scale becoming edged with black, giving a netted appearance. Unfortunately, in fish competitions the judges are quick to notice such artificial practices, and colour-fed fishes are likely to be down-graded for not complying with the natural colours of their species. The use of colour-enhancing lamps will also give the impression of more brightly coloured fishes, but naturally the fishes will regain their normal colours when removed to more normally lit environments using standard lamps in the hoods.

Below: *The sparkling colours of this Congo Tetra are produced by light reflecting from the guanin crystals just beneath the fish's skin. Many iridescent fishes look their best when illuminated with side-lighting.*

Popular Community Fishes

Fishes described in this part of the Guide are grouped according to their natural swimming areas, ie surface and upper levels, mid-water and bottom. Because of this classification, fish families may occasionally be represented in more than one group, for instance the Hatchetfishes (*Carnegiella, Gasteropelecus*) are part of the Characin group, whose other members frequent different water levels in the aquarium.

Of course, the fishkeeper should not expect his chosen species to restrict themselves totally to the areas in which they have been placed – even the bottom-dwelling *Corydoras* will make an occasional visit to the water surface, and the live-bearing fishes, although admirably suited for surface feeding with their upturned mouths, will nevertheless be among the first in the race to take food from the aquarium floor.

Sixty species are described, which should give sufficient range of choice, even allowing for personal tastes and local availability. Where only one species is described, eg *Julidochromis*, it may be assumed that other members of the genus share the same habits.

At breeding time, some species will of necessity change zones, eg the Gouramies spend much more time at the water surface under their floating bubble-nests.

The choice of genera shown in this Guide has been dictated by two factors: space, and suitability. You may be surprised (or even disappointed) at the omission of some species that are often featured fully in other aquarium books; even had there been room for some of them they would have been excluded on the grounds of their need for different aquarium conditions, no matter how well-behaved they are. To give an idea of scale, a line drawing of each species is compared with a silhouette of the Tiger Barb, which is 45mm (1.8in) in length, excluding the tail.

All the genera described (including related species) have been chosen for their ability to tolerate each other's company and similar aquarium conditions. Providing that the commonsense rule of avoiding extremes of size (in either direction) is observed, a carefully chosen selection of fishes from all three groups will provide you with an aquarium full of life and colour.

Surface swimmers

Most of the surface-swimming fishes are predatory to some degree, although the object of their attentions may vary in size from floating insects to smaller fishes. Their almost straight dorsal surface enables them to hang immediately below the water surface or under floating plant leaves waiting for food. When kept in the community tank they will have little opportunity to prey on fishes and must therefore be content with the varied diet that the fishkeeper provides.

Note: Lengths quoted are aquarium maxima, excluding tail.

Family: BELONTIIDAE
This group of fishes are particularly well-equipped to breathe atmospheric air at the surface of the water; an accessory breathing organ of a labyrinthine construction is situated behind the gills, thus giving this Family its popular collective name of Labyrinth Fishes.

Betta splendens

Siamese Fighting Fish

- **Habitat:** Southeast Asia
- **Length:** 60mm (2.4in)
- **Sex differences:** Males have more elongated fins than females.
- **Aquarium breeding:** Possible. Bubble-nest builder.

A solitary male specimen (or a male and two females) may be kept with success in the community tank. If males are kept together fights will occur; these contests of strength seem quite acceptable in the fish's native Thailand, where sums of money are wagered upon the outcome, but the aquarist will not be too interested in these battles.

The various colour strains seen of this species are the result of very carefully controlled breeding programmes carried out over the years by fishkeepers – the colours do not occur in nature, where their very brightness would be a threat to the fish's safety.

Like other Anabantid fishes, the Siamese Fighting Fish can breathe atmospheric air, which it makes use of in its labyrinth organ.

Below: **Betta splendens**
This red male spreads his impressive fins to deter possible rivals.

Family: CYPRINIDAE

A major contribution to the sheer liveliness of the aquarium is provided by the top-swimming fishes in this family group. Their slender, torpedo-shaped bodies indicate that they are fast swimmers, often just as at home in a mountain stream as in the slower moving waters of the jungle.

Unlike the Characidae, these fishes have no teeth in the jaws, but they do have teeth in the throat to grind up their food.

Brachydanio albolineatus

Pearl Danio

- **Habitat:** Southeast Asia
- **Length:** 55mm (2.2in)
- **Sex differences:** Female slightly stouter than the male, and less brightly coloured.
- **Aquarium breeding:** Possible. Egg-scatterer.

One of the main attractions of this species is the ever-changing delicate hues of colour on the flanks. These are best seen when some light shines through the front glass of the aquarium; when lit from above the effect is not seen so dramatically. It is a fish that revels in sunlight and its tank could be situated so that a little sun strikes it each day. (Remember, though, that too much direct sunshine can cause overheating and unsightly growths of algae.)

Breeding follows the standard method for egg-scatterers, and some precaution should be taken to prevent egg-eating by the adult fishes once spawning is over.

Below: **Brachydanio albolineatus**
The streamlined flanks are marked by lines which catch the light.

Brachydanio rerio

Zebra Danio

- **Habitat:** Eastern India
- **Length:** 50mm (2in)
- **Sex differences:** Females deeper in the body, especially when ready to breed. Males may have a flatter dorsal surface.
- **Aquarium breeding:** Possible. Egg-scatterer.

An easily recognized species, more than aptly described by its common name. A very active fish, always on the move. There is some speculation that the ground colour between the stripes is either gold or silver; at the speed at which the fish swims it is very difficult to decide.

This fish is almost unanimously voted 'beginner's breeding fish' due to the ease with which it can be bred in the aquarium. A shoal of Zebras can be spawned collectively, following the customary separate conditioning of the sexes.

Below: **Brachydanio rerio**
These fishes are always on the move in the tank and are easy to spawn.

Danio aequipinnatus

Giant Danio

- **Habitat:** Southwestern India and Sri Lanka.
- **Length:** 100mm (4in)
- **Sex differences:** As for *Brachydanio albolineatus*
- **Aquarium breeding:** Possible. Egg-scatterer.

Left: **Danio aequipinnatus**
A shoal of these fishes needs an aquarium with plenty of swimming room. This is a very active species.

The largest of the genus and a very handsome fish. The body is deeper and altogether chunkier than the smaller Danios, and is highly irridescent under side-lighting. It is a very prolific fish and will lay over 1000 eggs at a spawning. This striking and undemanding fish looks best when kept in a shoal.

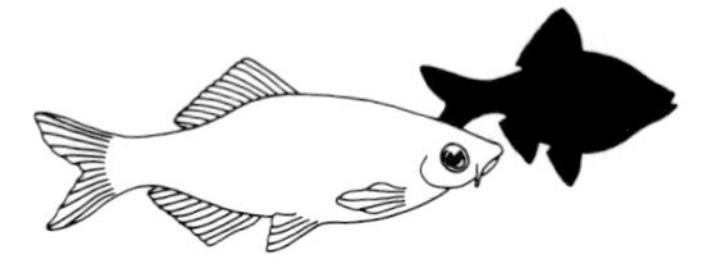

Laubuca dadiburjori

Dadio

- **Habitat:** Southeast Asia
- **Length:** 45mm (1.8in)
- **Sex differences:** Females probably plumper than the males at breeding time.
- **Aquarium breeding:** Possible. Egg-scatterers (as reported by hobbyists), although they may be egg-depositors in nature.

A fish that is not accorded the attention it deserves. Another species that must be seen in daylight, preferably in the morning sunshine, when its electric-blue lines can be seen to perfection.

Breeding has been seen to occur in clumps of plants and in puddles of water on the surface of floating leaves of Indian Fern (*Ceratopteris thalictroides*) in the aquarium.

These fish are expert at jumping, so their aquarium should be well-covered to prevent escape.

Below: **Laubuca dadiburjori**
The blue line on the side is best seen when sunlight hits the tank.

Rasbora dorsiocellata

Eye-spot Rasbora

- **Habitat:** Malaysia
- **Length:** 65mm (2.6in)
- **Sex differences:** Females are slighly plumper than males.
- **Aquarium breeding:** Possible. Egg-scatterer.

A silvery fish, with a dark blotch in the centre of the dorsal fin. A smaller subspecies *R. dorsiocellata macrophthalma* is similarly coloured but has a brilliant blue-green iridescent zone beneath its relatively larger eyes.

Right: **Rasbora dorsiocellata**
This elegant and normally prolific fish will thrive in slightly soft water.

Tanichthys albonubes

White Cloud Mountain Minnow

- **Habitat:** Southeast Asia
- **Length:** 45mm (1.8in)
- **Sex differences:** Males more brightly coloured perhaps.
- **Aquarium breeding:** Possible. Egg-scatterer.

Said to have been discovered by a Chinese Boy Scout, Tan (Tanichthys = Tan's Fish), this fish could be mistaken for a Neon Tetra when young, such is its brilliant colouration. A fish that does not need high temperatures and can be kept outdoors in summer months. Undemanding and easy to breed.

A similar looking fish was once described as a new species but it is only a colour variety. Recently a long-finned version of this lovely species has been developed.

Below: **Tanichthys albonubes**
This undemanding fish is often described as the 'poor man's Neon'.

Family: CYPRINIDONTIDAE

The fishes in this group are often known by either of two collective common names, 'Egg-laying Toothcarps' or 'Killifishes'; the second part of the former name refers to the fact that these fishes have teeth in the mouth, the latter is derived from a local word 'killi' meaning ditch. Most of the species are elongate and lurk just beneath the surface of the water among floating plants, from where they can dash out to capture their prey – either a smaller fish or an insect floating on the water surface.

The most remarkable asset is the ability of the fertilized eggs to withstand prolonged periods of drought; in nature, the fishes' habitat often dries out completely, so many species bury their eggs in the substrate of the stream bed. The eggs hatch when the rainy season refills the stream.

Above: **Oryzias melastigma**
This fish inhabits flooded rice paddy fields in its native regions.

Oryzias melastigma

- **Habitat:** India, Sri Lanka
- **Length:** 45mm (1.8in)
- **Sex differences:** Female may be slightly larger than the male.
- **Aquarium breeding:** Possible. Method of reproduction unusual.

An obvious surface swimmer from its appearance. This fish can tolerate reasonably low temperatures, and also some varying water conditions ranging from fresh to brackish water.

However, fishkeepers are attracted to this fish by its breeding methods. The eggs are fertilized by the male during a nuptial embrace, during which the male encloses the female body in his large anal fin. The fertilized eggs remain attached to the female for some time (almost like a miniature bunch of grapes) until they are brushed off by plants as she swims about. The onset of breeding is often brought about by a change of water conditions – the addition of some fresher water to the aquarium will often suffice.

Oryzias is a shoaling fish by nature and ideally should be kept in sufficient numbers for it to feel secure in the aquarium.

Pachypanchax playfairii

Golden Panchax; Playfair's Panchax

- **Habitat:** Zanzibar, Seychelles
- **Length:** 90mm (3.5in)
- **Sex differences:** Female has a dark blotch on the base of the dorsal fin.
- **Aquarium breeding:** Possible. Egg-scatterer.

This colourful miniature pike-like fish (whose aggressive tendencies it seems to share) is an established aquarium favourite.

Many fishkeepers start to worry when the scales on this fish rise up, exhibiting what is often taken as the symptoms of the disease 'dropsy'; this characteristic seems to occur mainly at spawning time and is nothing to be alarmed about.

Breeding takes place among clumps of plants over a period of days. A thick layer of the floating plant *Riccia* will provide a good spawning medium for the fishes and a perfect refuge for the young. The young grow quickly after hatching and can be netted out and transferred to another aquarium before their predatory parents catch up with them and eat them!

Above: **Pachypanchax playfairii**
This Killifish should be kept in an aquarium with similar-sized fishes.

Family: GASTEROPELECIDAE
The two species described here are equipped with certain physical characteristics that enable them to follow their particular life style. It is an excellent example of evolutionary development.

Carnegiella marthae

Black-winged Hatchetfish

- **Habitat:** South America
- **Length:** 45mm (1.8in)
- **Sex differences:** Female slightly plumper when viewed from above or from the front.
- **Aquarium breeding:** Possible. Egg-scattering among roots of floating plants.

The deep body of this species contains powerful pectoral muscles which by flapping the pectoral fins enable the fish to 'fly' through the air for some distance. Needless to say, the aquarium should be well-covered to prevent escape. Food is taken from the surface; food falling through the water is often ignored.

Below: **Carnegiella marthae**
The body outline, with a flat dorsal surface, indicates a top swimmer.

Gasteropelecus levis

Silver Hatchetfish

- **Habitat:** South America
- **Length:** 65mm (2.6in)
- **Sex differences:** As for *C. marthae.*
- **Aquarium breeding:** Not yet.

Members of the *Gasteropelecus* genus may be distinguished by the presence of an adipose fin – a small extra fin between the dorsal fin and the caudal fin. Like *C. marthae*, this species is often regarded as delicate by some fishkeepers.

Above: **Gasteropelecus levis**
These Hatchetfishes relish floating insect food at the water surface.

Family: LEBIASINIDAE
Members of this Family are also often referred to as belonging to the Characidae (in fact the Lebiasinidae and the Characidae are both members of the Order Cypriniformes within the Sub-Order Characoidei). The use of the words 'Tetra' or 'Characin' in these fishes' common names only prolongs this misconception, although the exactitudes of scientific classification does nothing to mar their attraction for the hobbyist, nor do the fishes seem to mind what name they are popularly known by!

Copella arnoldi (formerly Copeina)

Splashing, or Spraying, Tetra

- **Habitat:** South America
- **Length:** 75mm (3in)
- **Sex differences:** Males have more colour, more developed finnage. Females may be deeper and plumper in the body.
- **Aquarium breeding:** Possible. Egg-depositor, laying eggs out of water on overhanging leaves.

Above: **Copella arnoldi**
A pair of Splashing Tetras spawning on a leaf overhanging the water.

This fish is usually kept out of the interest generated by its peculiar breeding manner, as fishes that breed out of water are not all that common! In nature, the female lays the eggs on the underside of a leaf overhanging the surface of the water and the male, by an equally energetic leap, fertilizes them. To prevent the eggs from drying out before hatching occurs, the male constantly splashes them with water with deft flicks of his tail. As the young hatch they fall into the water and develop normally.

Recent literature has suggested that this species may be *C. compta* or *C. vilmae* and not *C. arnoldi* as described by Regan in 1912.

Family: PANTODONTIDAE
This Family from tropical West Africa contains a single species, which has a certain fascination for some fishkeepers.

Pantodon buchholzi

Butterfly Fish

- **Habitat:** West Africa
- **Length:** 110mm (4.3in)
- **Sex differences:** Rear edge of anal fin is straight in the female fish but concave in the male.
- **Aquarium breeding:** Possible, but not achieved with any regularity.

A fish for the fishkeeper who wants something 'different'. Like the Hatchetfishes, the Butterfly Fish can glide across the surface of the water using its wide pectoral fins as wings, although it does not beat its fins.

Observers of its breeding behaviour report that after vigorous male driving the fishes ride 'piggy-back' during the spawning act.

Butterfly Fishes appreciate some floating plants under which to lurk, or plants that throw up leaves to the water surface. Why Butterfly Fish? Try looking at them from above and you will see why – the large pectoral fins are usually patterned. A final note: the Butterfly Fish can be predacious towards smaller fishes.

Above: **Pantodon buchholzi**
Although rather drab in colour, this is a fascinating fish to keep.

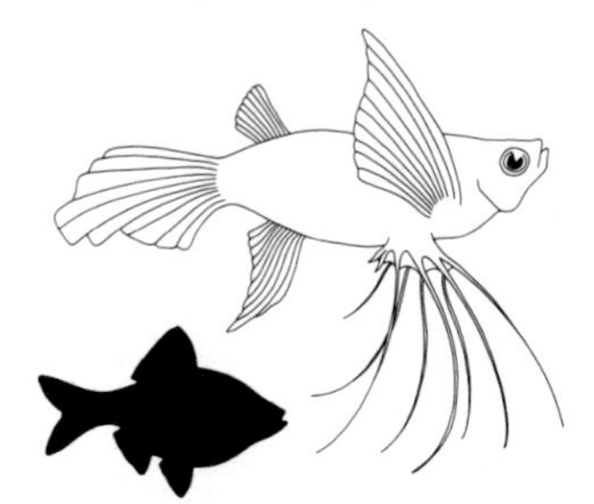

Family: SILURIDAE

This Family contains the true Catfishes. The species described here may be considered unusual in that it frequents the upper layers of the water, unlike its other relatives who are generally expected to be found on the stream bottom, where they use their sensitive barbels to orientate and search for food.

Kryptopterus bicirrhus

Glass Catfish

- **Habitat:** Indo-China and the Sunda Islands
- **Length:** 90mm (3.5in)
- **Sex differences:** Unknown.
- **Aquarium breeding:** Not yet achieved.

The transparency of the body is the immediate impression upon the viewer; other notable features are the single-rayed dorsal fin, the long pectoral fins and the oblique swimming attitude of the fish. Like *Oryzias*, this is a fish that feels most comfortable when in the company

of its own kind and it should never be kept alone, for it would soon pine away.

The Glass Catfish swims and rests at an oblique angle, with its caudal fin constantly waving from side to side. The species can be very colourful, despite its transparency; the body is highly irridescent and the colours seen depend upon the angle and intensity of the light. Unlike some of its relatives, this species is quite happy to swim during daylight.

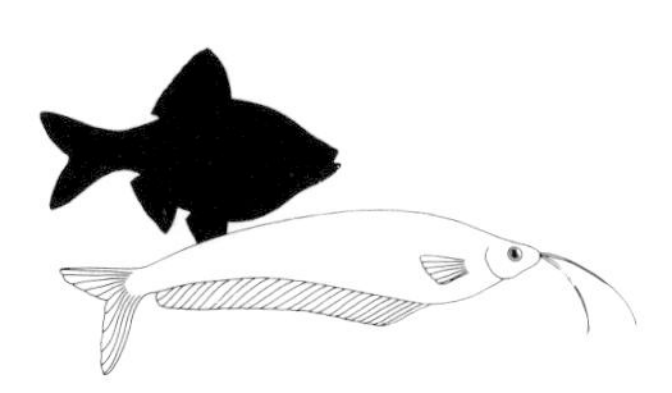

Below: **Kryptopterus bicirrhus**
These fishes are often kept for their unusual form and lack of 'body'.

Mid-water swimmers

Fishes that frequent the mid-water areas are deeper bodied generally, and their profile is symmetrical – neither the dorsal nor the ventral outline is flattened as it is in the case of surface- and bottom-swimming fishes respectively. The depth of the body is more suited to the slower moving waters, where the pressure on the fishes' flanks from the water flow is not as high as it would be in swifter flowing streams.

The pace of life is more leisurely and the Angelfish, with its disc-shaped body, is a good example of this. However, life is not all peace; many of the mid-water swimmers are carnivores and their terminally-situated mouths are filled with sharp teeth.

Breeding patterns are egg-scattering and depositing, nest-building and live-bearing.

Family: BELONTIIDAE

The graceful and beautiful Gouramies (also air-breathing Labyrinth Fishes like the Siamese Fighting Fish) are popular with fishkeepers; they investigate their territory with elongated pelvic fins that have taste buds at their tips. These 'feelers' may be as long as the fish's body height.

Colisa chuna

Honey Gourami

- **Habitat:** Northeastern India
- **Length:** 45mm (1.8in)
- **Sex differences:** Male more colourful, with yellow dorsal fin and deep turquoise throat at breeding time; female drab with a horizontal brown line running along the body.
- **Aquarium breeding:** Possible. Bubble-nest builder.

A relatively recent addition to the aquarium, the Honey Gourami has quickly become a favourite with aquarists the world over.

Like most of the genus, the male takes breeding very seriously and becomes quite aggressive for such a small fish. Despite their small size, the fry are not difficult to raise, but the first food for the fry must be of microscopic proportions.

Below: **Colisa chuna**
The male shown here is startling in colour compared to the drab female.

Colisa sota

Dwarf Gourami

- **Habitat:** Northeastern India
- **Length:** 60mm (2.4in)
- **Sex differences:** Male more colourful and with pointed fins; female silvery.
- **Aquarium breeding:** Possible. Bubble-nest builder.

A very colourful fish, with alternating bands of red and metallic blue sloping across the flanks. Colours are intensified at breeding time, when the male can become quite aggressive in nature.

Spawning begins with the building of the bubble-nest by the male fish, often using pieces of aquarium plants in the process. The female is then enticed beneath the nest, where spawning occurs. The fishkeeper should provide plenty of plants for the female to seek refuge in as the male will harrass her even after spawning. He will also attack her before spawning begins if he considers that she is not worthy of his amorous attentions!

The young fry are often found to be more difficult to raise than their 'cousins' the Honey Gourami. The aquarium temperature should be kept fairly high – above the normal 24°C (75°F) – for breeding.

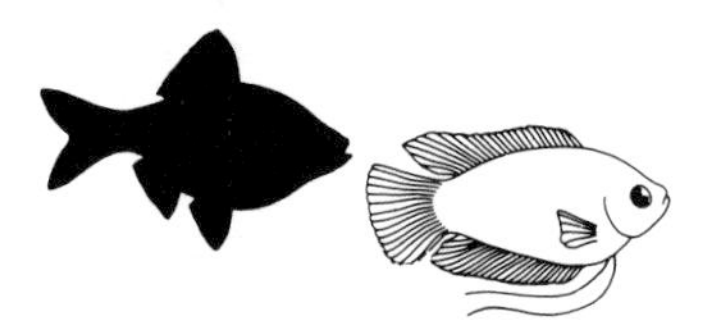

Right: **Colisa sota**
If your male fish has unbroken lines across its flanks, it's a winner!

Above: **Trichogaster leeri**
This male has a long dorsal fin.

Trichogaster leeri

Lace Gourami; Leeri Gourami; Mosaic Gourami

- **Habitat:** Far East
- **Length:** 110mm (4.3in)
- **Sex differences:** The male has a much longer dorsal fin (the female's is rounded) and will develop a fiery orange/red colour to the throat region and pectoral and anal fins during breeding.
- **Aquarium breeding:** Possible. Bubble-nest builder.

The pattern on the sides of this fish forms a mosaic or lacy design and a dark line runs through the eye across the flanks to the caudal peduncle. Adult males develop filamentous extensions to their dorsal and anal fins.

Fishkeepers wanting to breed this species find that they usually have a reasonable time to wait; this fish appears to take a long time to become sexually mature.

Trichogaster trichopterus

Blue Gourami; Three-spot Gourami

- **Habitat:** Far East
- **Length:** 110mm (4.3in)
- **Sex differences:** The male's dorsal fin is more pointed than the female's.
- **Aquarium breeding:** Possible. Bubble-nest builder.

This pale blue fish with three spots (the third spot is the eye) is a peaceful and prolific species. The males are very vigorous drivers at breeding time; remove the female after spawning is completed.

Above: **Trichogaster trichopterus**
A female with a short dorsal fin.

A subspecies, *T. trichopterus sumatranus* (the Opaline Gourami) has dark wavy or mottled lines on the flanks; a recent introduction of a colour variation of this subspecies is the Gold Opaline Gourami.

Family: CALLICHTHYIDAE

Members of this Family are generally known as the Armoured Catfishes, and this prompts the question what a Catfish is doing swimming around in mid-water! The single species from the genus *Dianema* described here is one of those exceptions to the rule, and generally spends most of its time away from the floor of the aquarium.

Dianema urostriata

Striped-tailed Catfish; Flag-tailed Catfish

- **Habitat:** Brazil
- **Length:** 130mm (5.1in)
- **Sex differences:** Unknown.
- **Aquarium breeding:** Apparently breeding has been achieved but the details are unknown, except that this species is a bubble-nest builder.

A strikingly good-looking fish with a black-and-white striped caudal fin. In the aquarium it may tend to be on the shy side, hiding away during the hours of 'tanklight'; it is also a gregarious species and therefore should be kept in a shoal.

Like the related *Corydoras* Catfishes, *Dianema* has armoured plates but far longer barbels.

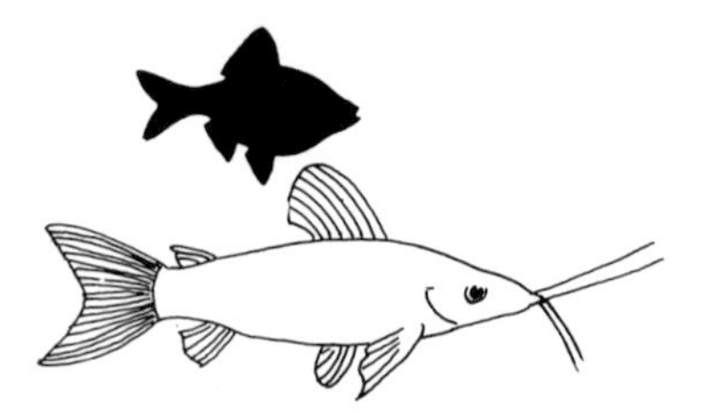

Below: **Dianema urostriata**
This usually shy fish is showing off its boldly striped caudal fin.

Family: CHARACIDAE
A very large Family of fishes containing such diverse species as the brilliantly coloured Neon Tetra and the dreaded Piranha. The majority of species are excellent aquarium fishes with much to commend them – colour, finnage and, in general, a peaceful disposition. Most are hardy and healthy eaters but a few may need special water conditions or take an unhealthy interest in the aquarium plants!

Paracheirodon axelrodi

Cardinal Tetra

- **Habitat:** South America
- **Length:** 45mm (1.8in)
- **Sex differences:** Males have tiny plates on the underside of the caudal peduncle; females may be slightly larger.
- **Aquarium breeding:** Possible. Egg-scatterer.

A truly gorgeous fish with intense colours. Best seen in a shoal. It requires soft water in which to breed (acclimatize the fish to this new water condition slowly) and spawning often occurs in dim light. Some fishkeepers keep a very low wattage lamp burning over the aquarium to simulate the optimum conditions for spawning.

Below: **Paracheirodon axelrodi**
A shoal of these attractive fishes provides a magnificent spectacle.

Gymnocorymbus ternetzi

Black Widow; Black Tetra; Petticoat Fish; Blackamoor

- **Habitat:** South America
- **length:** 55mm (2.2in)
- **Sex differences:** Females are slightly larger than the males.
- **Aquarium breeding:** Possible. Egg-scatterer.

This is a very popular Tetra as its dark colour is an excellent contrast to the colours of most of the fishes kept in the aquarium. Unfortunately, the jet-black fades to dusky grey as the fish matures. An aquarium-developed, long-finned variety has been introduced in recent years. The species' black, fan-shaped silhouette is best seen when the fishes are kept in a shoal.

They are easy to breed, and the male's courting display has been described as a fluttering dance before the female of his choice.

Right: **Gymnocorymbus ternetzi**
A good all-round fish, tolerant of others and generally easy to breed.

Above:
Hemigrammus erythrozonus
Despite its lack of gaudy colours, the Glowlight Tetra is still a favourite.

Hemigrammus erythrozonus

Glowlight Tetra

- **Habitat:** South America
- **Length:** 45mm (1.8in)
- **Sex differences:** Males have very hard-to-see hooks on their anal fin; females are usually fatter at spawning time.
- **Aquarium breeding:** Possible. Egg-scatterer.

A more subtly coloured fish than the Cardinal Tetra, and, again, one that looks best when kept in a shoal.

When spawning this fish it is always advisable to use soft, slightly acid water to achieve best results.

Hemigrammus ocellifer

Beacon Fish

- **Habitat:** South America
- **Length:** 45mm (1.8in)
- **Sex differences:** The swim-bladder is more visible in the male, but otherwise the usual rule of 'fatter females' applies.
- **Aquarium breeding:** Possible. Egg-scatterer.

The common name of this species is prompted by the bright spot at the top of the caudal peduncle, which shines out against a dark backing.

Above: **Hemigrammus ocellifer**
Note the bright red eye, which almost matches the spot on the tail.

Hemigrammus rhodostomus

Rummy-nosed Tetra; Red-nosed Tetra

- **Habitat:** South America
- **Length:** 55mm (2.2in)
- **Sex differences:** Females are more robust.
- **Aquarium breeding:** Possible. Egg-scatterer.

An easily identifiable species with a red cap and striped caudal fin. A larger species (*Petitella georgiae*) is almost identical except for a slight difference in the area of red and the fact that it lays fewer eggs.

Below:
Hemigrammus rhodostomus
This fish needs to be in good health for its colours to be really vivid. In common with most Tetras, this species thrives in soft water.

Hyphessobrycon erythrostigma

Bleeding Heart Tetra

- **Habitat:** South America
- **Length:** 70mm (2.8in)
- **Sex differences:** The male has large sickle-shaped dorsal and anal fins.
- **Aquarium breeding:** Possible but not to any regular extent. Egg-scatterer.

The blood-red blotch on the side of this species (just behind the gill cover) gives this fish its common name. It usually lives for a long time in the aquarium, although some fishkeepers have found it to be of a rather nervous disposition when transferred from tank to tank during exhibitions and fish shows.

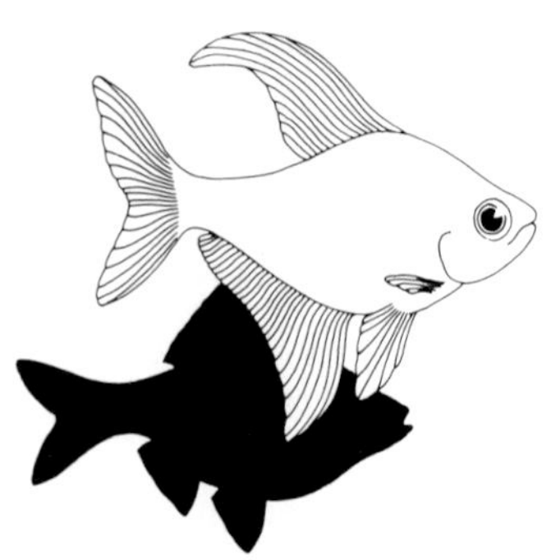

Right:
Hyphessobrycon erythrostigma
A fairly recent addition to the aquarium, this large Tetra has found favour with fishkeepers everywhere.

Hyphessobrycon pulchripinnis

Lemon Tetra

- **Habitat:** South America
- **Length:** 50mm (2in)
- **Sex differences:** As for *Hemigrammus erythrozonus.*
- **Aquarium breeding:** Possible. Egg-scatterer.

An established aquarium favourite. Its delicate lemon-yellow colours are best seen in the aquarium with a dark floor covering or against a background of wood or cork bark. The red colouring of the eye is quite startling. Recently, an albino version of this species has been introduced but it is an aquarium-made fish, not a naturally occurring species.

Below:
Hyphessobrycon pulchripinnis
The tiny adipose fin (often seen in the Characin group) is visible here between dorsal and caudal fin.

Megalamphodus megalopterus

Black Phantom Tetra

- **Habitat:** South America
- **Length:** 40mm (1.6in)
- **Sex differences:** Observe the usual rule – the females are plumper. The male has more developed finnage.
- **Aquarium breeding:** Possible. Egg-scatterer.

Along with its red look-alike relative *M. sweglesi*, this fish brings a contrasting colour to the aquarium.

Right:
Megalamphodus megalopterus
The male uses his large dorsal fin to impress females or deter rival males.

Micralestes (Phenacogrammus) interruptus

Congo Tetra; Congo Salmon

- **Habitat:** Central Africa
- **Length:** 90mm (3.5in)
- **Sex differences:** Males are more highly coloured and have a lengthened dorsal fin and filamentous extensions to the caudal fin.
- **Aquarium breeding:** Possible. Egg-scatterer.

This African Characin is best seen with a little side-lighting picking up the iridescence from the scales. This is a superb shoaling fish – an aquarium full of these fishes swimming among submerged wood logs cannot be bettered.

Left: **Micralestes interruptus**
Only the male fishes have the ornate extensions to their dorsal and caudal fins; these develop fully in a spacious aquarium with plenty of swimming room. These fishes look their best against a dark floor covering.

Moenkhausia pittieri

Diamond Tetra

- **Habitat:** South America
- **Length:** 65mm (2.6in)
- **Sex differences:** The male has more flowing fins than the female.
- **Aquarium breeding:** Possible. Egg-scatterer.

Although this species does not have the bright colours of the preceding one, it can match it glitter for glitter. Like the Congo Tetra, it has fins outlined in white which are best seen against a dark background. The eye has a bright red spot.

Below: **Moenkhausia pittieri**
This fish gained its popular name of Diamond Tetra from the sparkling display of its iridescent body.

Nematobrycon palmeri

Emperor Tetra

- **Habitat:** South America
- **Length:** 60mm (2.4in)
- **Sex differences:** The male has a long extension to the dorsal fin and a long centre spike to the caudal fin.
- **Aquarium breeding:** Possible. Egg-scatterer.

A beautiful blue and black fish with bright yellow fins. The eye is a lovely turquoise in colour.

Although it will breed in the aquarium, it is not a very prolific fish.

Below: **Nematobrycon palmeri**
The Emperor Tetra rivals any tropical fish, with its impressive range of body colours and blue-green eye.

Paracheirodon innesi

Neon Tetra

- **Habitat:** South America
- **Length:** 42mm (1.7in)
- **Sex differences:** Females are plumper and often best recognized by the 'bend' in the electric-blue line along the flanks.
- **Aquarium breeding:** Possible. Egg-scatterer.

Until the much more gaudy Cardinal Tetra came along, this was the jewel of the aquarium according to most people. Like some other species, this fish is often recognized by non-fishkeeping people (who are unaware of its scientific relationship to the Piranha!). Many newcomers make the mistake of keeping this species with fish that will ultimately grow larger than the Neons, with the predictable result – no Neons! When buying fish, always try to find out their eventual fully-grown size.

Left: **Paracheirodon innesi**
This colourful and hardy species was responsible for a marked increase in fishkeeping during the 1930s. It is now a universally popular species.

Family: CICHLIDAE

The Cichlid fishes are amongst the most popular with fishkeepers. The appeal is not so much their visual beauty (with some species, even their mothers wouldn't call them good-looking!), but their behaviour at breeding time, when parental care is seen at its best. Unfortunately, the price to pay for the privilege of seeing this display includes having to tolerate some pretty unsociable behaviour from the fishes concerned, as many are aggressively territorial, especially at breeding time. For this reason, only a few Cichlids can be recommended for inclusion in the community aquarium.

Aequidens curviceps

Sheepshead Acara; Flag Cichlid

- **Habitat:** South America
- **Length:** 75mm (3in)
- **Sex differences:** The male has more pointed fins than the female.
- **Aquarium breeding:** Possible. Egg-depositor.

A reasonably peaceful fish that lays its eggs on a stone in full view of the fishkeeper. About 200 eggs are laid and these hatch in three days. Like most Cichlids, this species appreciates clean tank conditions, and frequent partial water changes should be the rule rather than the exception for success.

Above: **Aequidens curviceps**
This small shy Cichlid will breed readily in the community aquarium. Watch out for it in the dealer's tanks – it can be quite a rarity.

Pterophyllum scalare

Angelfish

- **Habitat:** South America
- **Length:** 100mm (4in)
- **Sex differences:** Observation of the spawning tubes extended by the fishes at breeding times is the only accurate method, despite many 'recipes' appearing in aquatic books. Another clue is that if two fishes 'pair up' and keep each other's company constantly then they may be a true pair – male and female – but until one fish lays the eggs you still won't know for sure which sex is which. One consolation is that sometimes even the fishes themselves can't tell which is which either!
- **Aquarium breeding:** Possible. Egg-depositor.

This fish hardly needs any introduction as it is so well-known. Since it rapidly became an aquarium favourite many years ago, aquarium breeding has developed many new colour and finnage strains in addition to the original silver; today there are Black, Marbled, Zebra and Gold Angels, to name but four.

Angelfishes are very hardy and are long lived. They are generally peaceful and will not uproot plants when marking out a territory. However, once they become large they may take a fancy to small fishes present in the community aquarium.

Sometimes the adult pair of fishes refuse to raise their young and start to eat them after a few days. In this case, once the eggs have been laid and fertilized, the fish should be removed and the eggs hatched artificially with a well-placed airstone taking over the function of the parents' fin-fanning actions. Once the young fishes begin to swim freely (after about ten days) they can be fed on the usual first foods, such as rotifers and nauplii.

Below: **Pterophyllum scalare**
Like the Neon Tetra, this superb fish has become a firm favourite; the many colour strains available today underline its continuing popularity.

Family: CITHARINIDAE
This Family is the African counterpart to the South American Characidae and contains both large and small species.

Nannaethiops unitaeniatus

One-striped African Characin

- **Habitat:** Equatorial Africa
- **Length:** 65mm (2.6in)
- **Sex differences:** The female is deeper in the body; the male has a red dorsal and upper caudal fin at breeding times.
- **Aquarium breeding:** Possible. Egg-scatterer.

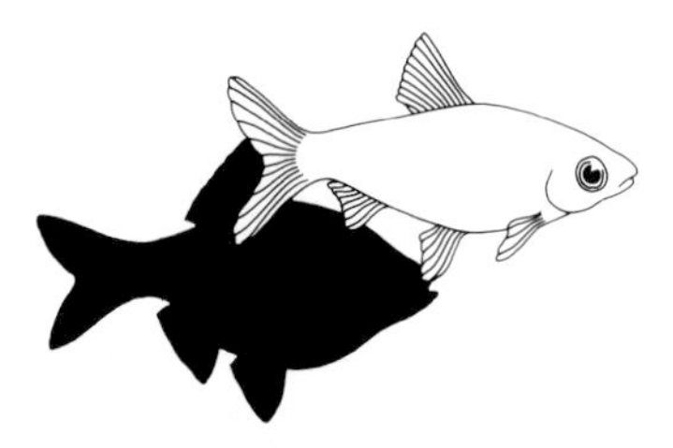

A peaceful fish which despite its rather drab appearance is a good, prolific community fish and can be bred easily in the aquarium.

Below: **Nannaethiops unitaeniatus**
Morning sunshine striking the tank can stimulate spawning in this very productive and hardy species.

Family: CURIMATIDAE:
Another American Family of Characins, often confusingly described as Barb Characins, which refers to their body form as being more like Barbs than Characins. Many of the species within this group display a 'head-down' attitude when resting or swimming and the common name 'Headstanders' reflects this characteristic.

Chilodus punctatus

Spotted Headstander

- **Habitat:** South America
- **Length:** 80mm (3.2in)
- **Sex differences:** The female is plumper than the male.
- **Aquarium breeding:** Possible. Egg-scatterer. There may be problems in getting the young fish out from the eggs, and various ways (including the use of a scalpel) have been described by eminent aquarists.

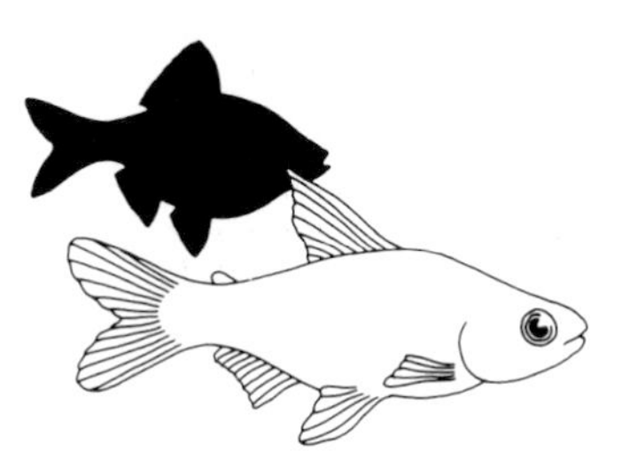

An easily recognizable fish, not only from its head-standing attitude but also from the spotted scales and large squared-off dorsal fin.

Below: **Chilodus punctatus**
The characteristic oblique swimming position of the Headstanders may help them to hide among vegetation.

Family: CYPRINIDAE

A very large Family of fishes which are distributed over much of the world, with the notable exceptions of South America, Malagasy and Australia. They include the Carps and Carp-like fishes and make up a very large part of the tropical freshwater aquarium scene. Some of the Barbs (*Barbus* spp) have a reputation for nipping other, slower-moving fishes' fins, but this tendency may be due to their being kept as isolated specimens and does not occur so much if the fishes are kept in sufficient numbers.

Barbus conchonius

Rosy Barb

- **Habitat:** Northeastern India
- **Length:** 75mm (3in)
- **Sex differences:** Males have more colour to their fins (pink and black) and turn a very deep rosy red colour at breeding times. Females are noticeably plumper.
- **Aquarium breeding:** Possible. Egg-scatterer.

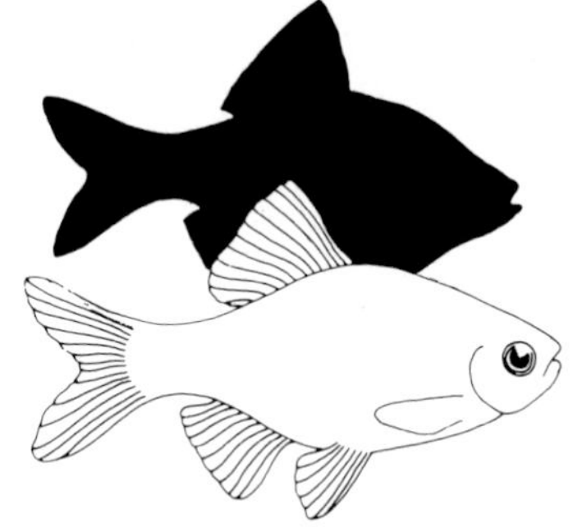

A very prolific fish which can be easily bred. Recently, a long-finned variety of this undemanding fish has been produced by selective aquarium breeding programmes.

Below: **Barbus conchonius**
This is a very popular Barb and an excellent choice for beginners. It will thrive in subdued lighting and breed readily in the community aquarium.

Above: **Barbus cumingi**
A lively and attractive fish from Sri Lankan rivers, where two different fin-colour varieties exist.

Barbus cumingi

Cuming's Barb

- **Habitat:** Sri Lanka
- **Length:** 50mm (2in)
- **Sex differences:** The male has deeper coloured fins; the female may be fractionally larger.
- **Aquarium breeding:** Possible. Egg-scatterer.

An active fish which should be kept in a shoal. It likes to include some vegetation in its diet so the aquarium plants should be tough enough to withstand a little nibbling. Soft water is advantageous when breeding but the adults may be kept in hard water without any difficulty.

Barbus nigrofasciatus

Purple-, or Ruby-headed Barb

- **Habitat:** Sri Lanka
- **Length:** 60mm (2.4in)
- **Sex differences:** No trouble at all here, the male is much darker with little silver showing between the bands on the flanks. At breeding time the male's head region turns a deep rich ruby red.
- **Aquarium breeding:** Possible. Egg-scatterer.

An established aquarium favourite. Although quite hardy, some fishkeepers use this fish as a barometer of conditions in their tank; if there is any hint of white spot disease about this Barb usually contracts it first! Fortunately, it responds to the proprietary treatments very readily.

Below: **Barbus nigrofasciatus**
This is a young female of the species; note the colourless fins.

Barbus oligolepis

Checker Barb; Island Barb

- **Habitat:** Sumatra
- **Length:** 45mm (1.8in)
- **Sex differences:** The male is more colourful and his single fins have black edges. Females are deeper in the body.
- **Aquarium breeding:** Possible. Egg-scatterer.

Right: **Barbus oligolepis**
The electronic flash highlights the iridescent colours of the fins.

The patterns on the scales give this fish a checkerboard appearance. This Barb is particularly suitable for the smaller aquarium, where a shoal can form the basis of an attractive community collection. It will breed readily, especially if the tank water is mature and on the soft side.

Barbus titteya

Cherry Barb

- **Habitat:** Sri Lanka
- **Length:** 45mm (1.8in)
- **Sex differences:** The male is more red in colour; the female is brown.
- **Aquarium breeding:** Possible. Egg-scatterer.

A somewhat shy fish, but quite suitable for the community aquarium, where it should be kept in reasonable numbers. The male takes on glowing red colours at spawning time. The species is not as prolific as some Barbs.

Below: **Barbus titteya**
This photograph clearly shows the barbels around the fish's mouth.

Rasbora borapetensis

Red-tailed Rasbora

- **Habitat:** Thailand
- **Length:** 55mm (2.2in)
- **Sex differences:** The female is deeper bodied than the male.
- **Aquarium breeding:** Possible. Egg-scatterer.

This is an elegant species with a black and gold band along the flanks and a delicate red hue to the tail. For successful spawning, keep the lights in the tank fairly dim.

Right: **Rasbora borapetensis**
These fishes, like the Danios, bring welcome activity to the aquarium.

Rasbora heteromorpha

Harlequin Fish

- **Habitat;** Malaysia, Thailand
- **Length:** 45mm (1.8in)
- **Sex differences:** The 'triangle' marking is more clearly defined with sharp points in the male fish.
- **Aquarium breeding:** Possible. Egg-depositor.

Another instantly recognizable fish, but this species differs from other Rasboras by laying its eggs on the underside of aquarium plant leaves (*Cryptocoryne* plants are useful in this respect) rather than scattering them indiscriminately.

Left: **Rasbora heteromorpha**
Keep these silvery fishes in a shoal to brighten the community aquarium.

Family: LEBIASINIDAE

The Pencilfishes described here have been the subject of many taxonomic re-classifications since their introduction to the aquarium world. Not all Pencilfishes spawn in the same manner; some are egg-scatterers while others may deposit their eggs on plant leaves.

Nannostomus beckfordi

Brown, or Red Pencilfish

- **Habitat:** South America
- **Length:** 50mm (2in)
- **Sex differences:** The anal fin is lobe-shaped in the male but straight-edged in the female.
- **Aquarium breeding:** Possible. Egg-scatterer.

The males of the species are forever dancing around the aquarium, either impressing the females or having mock battles with other males in the shoal.

Breeding can present a few problems as the fishes are egg-eaters and only lay a few eggs at a time. Add to this that the eggs may not develop due to the tank water conditions, and you will see that the fishkeeper who spawns these fishes can be proud of his (and the fishes') achievement.

Like all Pencilfishes, this species has a night-time colour pattern that reverts to the illustrated colours as soon as the aquarium is illuminated.

Right: **Nannostomus beckfordi**
These Pencilfishes are shown in the 'normal' daytime colour pattern.

Nannostomus trifasciatus

Three-lined,or Three-banded Pencilfish

- **Habitat:** South America
- **Length:** 50mm (2in)
- **Sex differences:** As for *Nannostomus beckfordi.*
- **Aquarium breeding:** Possible but rarely achieved. Egg-scatterer.

The nocturnal colour pattern of this fish consists of slanting bars instead of horizontal stripes. It will thrive in an aquarium with dense vegetation, provided it has sufficient space for swimming and slightly soft water.

Left: **Nannostomus trifasciatus**
With its distinctive stripes and red fins, this is an elegant Pencilfish.

Family: POECILIIDAE

The fishes in this group are live-bearing – they produce free-swimming fry at birth. Females of the four species described here will produce successive broods from a single mating, but other live-bearing species (the Goodeidae group for example) require a separate mating for each brood.

Because of their ability to produce regular broods, the live-bearing fishes have been used by the fishkeeper to produce, through carefully controlled breeding programmes, many internationally recognized colour strains.

Another valuable asset that some live-bearers possess is an enormous appetite for insect larvae; this has been put to good use in tropical countries where malaria is a problem. Live-bearing fishes are imported to eat the malaria-carrying mosquito larvae in the waterborne stage.

In the aquarium, the young of live-bearers form a tasty snack for the other fishes. Liver-bearers therefore provide an easy way to turn dried foods into live foods – what a morbid thought for a fishkeeper!

Poecilia reticulata

Guppy; Millionsfish

- **Habitat:** Northern South America, Barbados, Trinidad.
- **Length:** Males – 30-40mm (1.2-1.6in); females – 65mm (2.6in)
- **Sex differences:** All male live-bearing fishes have the anal fin adapted in such a way that it is easily distinguished from the female's fan-shaped anal fin. The Guppy male's anal fin has a rodlike appearance and can be swivelled in almost any direction.
- **Aquarium breeding:** Just try to stop them! Live-bearing.

A very colourful and active species. Through the efforts of fishkeepers even the usually drab females have some colour to their finnage nowadays. The problem with prolific live-bearers is keeping the strain pure, as any male (of any colour pattern) will pursue any female relentlessly, causing a colour strain to degenerate into a collection of nondescript fishes.

The sexes of live-bearer young must be closely watched for during their development stage and segregated ruthlessly; any of doubtful sex should be kept in with the males. The best way to maintain a pure strain is to keep only one or two colour strains (in separate collections) and specialize in these rather than give yourself problems keeping a variety of strains.

Left: **Poecilia reticulata**
The superb Guppy needs almost no introduction, it is so well known.

Poecilia mexicanus (P. sphenops)

Black Molly

- **Habitat:** Southern USA
- **Length:** Male – 70mm (2.8in); female – 95mm (3.7in)
- **Sex differences:** As for general live-bearers.
- **Aquarium breeding:** Possible. Live-bearing.

A very popular fish because of its black velvety appearance; a Lyretail variety is also available.

Gravid (pregnant) female live-bearers are very sensitive, particularly near to the end of their four week gestation period; moving them too late to a nursery aquarium may often result in premature birth of fry. Breeding traps seem almost fatal to gravid Mollies, and many fishkeepers do not use them.

Mollies are very appreciative of two things – some green vegetable food and some salt (natural sea salt crystals, not packaged table salt) added to the aquarium water.

Above: **Poecilia mexicanus**
The Black Molly's colour adds a useful contrast to the aquarium.

Xiphophorus helleri

Swordtail

- **Habitat:** Mexico
- **Length:** Male – 95-105mm (3.7-4.1in); female – 100-110mm (4-4.3in)
- **Sex differences:** In addition to the usual anal fin structure, the male Swordtail has a swordlike extension to the lower edge of the caudal fin.
- **Aquarium breeding:** Possible. Live-bearing.

Like other live-bearers, the Swordtail has been developed from its original green form into many colour strains and with exaggerated fin forms. It, too, likes some greenstuff in its diet.

One feature of the Swordtail is its occasional tendency to change sex, or rather the female to take on male characteristics, ie, develop a short 'sword' on the tail.

Swordtails are prolific, and at one time they held the record for the number of fishes in a single brood – something in excess of 270!

Right: **Xiphophorus helleri**
The male fish of this elegant pair is easily recognizable by the swordlike extension to the caudal fin.

Xiphophorus maculatus

Platy

- **Habitat:** Mexico, Guatemala, Honduras
- **Length:** 40mm (1.6in)
- **Sex differences:** As for general live-bearers.
- **Aquarium breeding:** Possible. Live-bearing.

Left: **Xiphophorus maculatus**
This attractive Platy is the same aquarium-developed colour strain as the Swordtail shown above but, of course, the male fish has no 'sword'.

A compact stocky version of its near relative the Swordtail, the Platy has no sword but shares all the Swordtail's aquarium-developed colour strains. It is a compulsive nibbler of any green algae that forms around the aquarium.

Bottom swimmers

Many bottom-swimming fishes are introduced to fishkeepers as 'scavengers', which is unfortunate as they have a perfect right to live the way they do and not be relegated to this 'tidying-up' category at all.

Generally, things to look out for in this group of fishes are a flattened ventral surface (which helps them to stay in position by preventing the water flow from getting under them and lifting them upwards), an underslung, suitably placed mouth often bristling with barbels with which to sense out food, and maybe larger than expected eyes for use in the dark and muddy surroundings. Don't be too disappointed if you see little of these fishes during the daytime – they are often nocturnal by nature.

Family: CALLICHTHYIDAE
This Family is best represented by the *Corydoras* genus, which contains many species that have become aquarium favourites all over the world. They have bony plates, an adipose fin, and a varying number of barbels.

Corydoras aeneus

Bronze Corydoras

- **Habitat:** South America
- **Length:** 75mm (3in)
- **Sex differences:** When viewed from above, the female is plumper than the male just behind the pectoral fins.
- **Aquarium breeding:** Possible. Egg-depositor.

This is the classic 'Cat', with an endearing habit of rolling its eyes when at rest on the aquarium floor.

Corydoras are gregarious and should be kept in shoals. To prevent their barbels becoming worn away by persistent digging in the gravel for food there are two remedies: give them food especially for them (not leftovers from the fishes swimming above them) and make sure that the gravel is not too sharp-edged, as this will wear away their 'whiskers' even faster.

When breeding, the female carries the fertilized eggs to a selected site (often a broad-leaved plant or the aquarium glass) and places the eggs there to hatch.

Below: **Corydoras aeneus**
This is the typical resting pose adopted by the Corydoras Catfishes.

Corydoras julii

Leopard Catfish

- **Habitat:** South America
- **Length:** 60mm (2.4in)
- **Sex differences:** As for *Corydoras aeneus*
- **Aquarium breeding:** Possible. Egg-depositor.

This species has been confused with two other Corydoras species of very similar appearance – *C. leopardus* and *C. trilineatus* – but with experience fishkeepers will soon come to learn the subtle differences – the spotted, shorter head than the others and the greater length of the horizontal line.

Right: **Corydoras julii**
This is one of the most strikingly patterned of all the Corydoras Catfishes. It is an active fish and best kept in a small shoal.

Corydoras reticulatus

Reticulated Catfish

- **Habitat:** South America
- **Length:** 65mm (2.6in)
- **Sex differences:** As for *Corydoras aeneus*
- **Aquarium breeding:** Possible. Egg-depositor.

This Catfish only attains its smart reticulated pattern at maturity. Like all Corydoras, this species can make use of atmospheric air (taken during a lightning dash to the surface) in its gut, and this enables the fish to survive in polluted waters, where other fishes may perish.

Remove both parents from the tank once spawning is completed.

Left: **Corydoras reticulatus**
This attractive fish is resting on a leaf, often a favourite spawning place for Corydoras Catfishes in general.

Family: CICHLIDAE
Some of this normally mid-water-swimming group of fishes frequent the lower levels of the aquarium, and these include some of the colourful Dwarf Cichlids and rock-dwelling African Lake species.

Aequidens maronii

Keyhole Cichlid

- **Habitat:** South America
- **Length:** 100mm (4in)
- **Sex differences:** The fins are more pointed in the male.
- **Aquarium breeding:** Possible. Egg-depositor.

The common name is descriptive of the keyhole type of marking on the fish's flanks. This somewhat shy fish is generally seen in the lower levels of the aquarium, where it shelters among the plants and rocks.

Above: **Aequidens maronii**
This peaceful Cichlid will lay eggs on a previously cleaned rock.

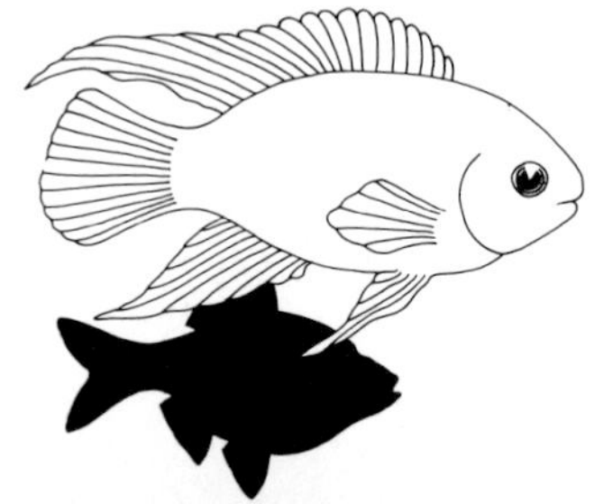

Apistogramma agassizi

Agassiz's Dwarf Cichlid

- **Habitat:** South America
- **Length:** Male 70mm (2.8in); female 57mm (2.2in)
- **Sex differences:** The male is easily recognizable, with a spade-shaped tail edged in white. The female is rather drab, with a dark brown line visible at times on the flanks. Females of different *Apistogramma* species are often difficult to identify – they are very similar in appearance.
- **Aquarium breeding:** Possible. Egg-depositor.

This fish, and other *Apistogramma* species, are secretive spawners, preferring to spawn away from the public gaze; usually a flowerpot laid on its side will be accepted as a spawning site quite readily, otherwise a rocky cave will be chosen. It is not unusual for the female to lay the eggs on the 'ceiling' of the flowerpot or cave. Dwarf Cichlids are quite territorially-minded towards other fishes at spawning times.

Below: **Apistogramma agassizi**
An excellent dwarf Cichlid that will not dig up the plants in the tank.

Thorichthys meeki

Firemouth Cichlid

- **Habitat:** South America
- **Length:** 150mm (6in)
- **Sex differences:** The male has more pointed fins and fin-rays than the female and develops a spectacular red colouring in the throat and ventral areas during spawning time.
- **Aquarium breeding:** Possible. Egg-depositors

Despite its size, this is a very peaceful Cichlid which keeps well out of the hurly-burly life of the aquarium. It is not likely to do much uprooting of plants, even during the spawning period, but the plants should be of a fairly robust nature (such as *Echinodorus* spp.) just to be on the safe side.

Left: **Thorichthys meeki**
This 'full frontal' aspect of the Firemouth deters any rival males.

Julidochromis ornatus

Golden Julie

- **Habitat:** Africa, Lake Tanganyika only
- **Length:** 90mm (3.5in)
- **Sex differences:** No external features.
- **Aquarium breeding:** Possible. Egg-depositor.

This torpedo-shaped fish is a cave-dweller and quite intolerant of its own kind; to keep a number of these in a small tank is asking for trouble and even in a large tank there must be plenty of hideaways available to avoid incessant squabbling between males. An isolated pair of *J. ornatus* may be kept successfully in a community tank providing there is some cave-like accommodation for them to seek refuge sometimes.

These fish are from a naturally hard water area and do well in areas where the tap water is similarly hard.

Above: **Julidochromis ornatus**
Easily identified by its bold stripes, this is an interesting hard-water Cichlid from Lake Tanganyika.

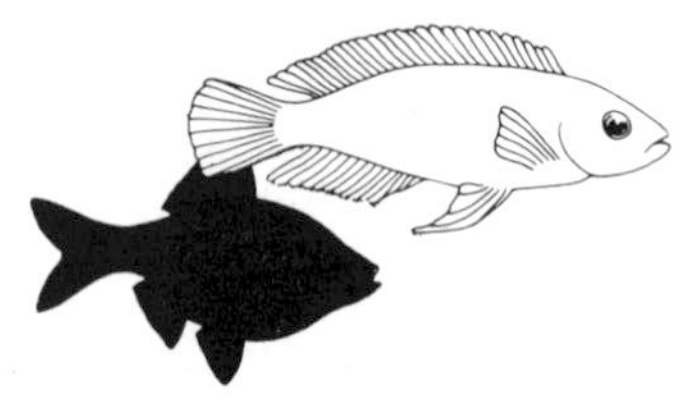

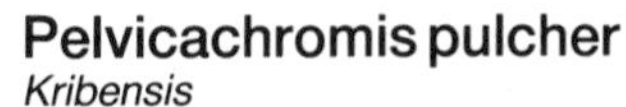

Pelvicachromis pulcher

Kribensis

- **Habitat:** West Africa
- **Length:** Male 100mm (4in); female 75mm (3in)
- **Sex differences:** The male has pointed fins; the female has dark dots in the rear end of the dorsal fin and has a deep plum colour at breeding time, and is often more brilliantly coloured than the male.
- **Aquarium breeding:** Possible. Egg-depositor.

Another secretive spawner; the fishkeeper only knows that breeding has occurred when the proud parents take their brood of young for an outing around the aquarium for the first time. Kribensis usually spawns beneath rocks or in flowerpots, or wherever the fishkeeper can't see what's going on! An established aquarium favourite, this fish was formerly called *Pelmatochromis kribensis.*

Below: **Pelvicachromis pulcher**
The Kribensis is usually a prolific fish and a conscientious parent.

Family: COBITIDAE
Members of this group, the Loaches, are popular (if difficult to net) fishes for the aquarium. Many possess spines which can be erected and these often get trapped in the fishkeeper's net. Like Corydoras they can also make use of atmospheric air in the hind gut. In nature, Loaches are often found in swiftly flowing streams, where they lurk behind or under stones.

Acanthophthalmus kuhli

Coolie Loach

- **Habitat:** Southeast Asia
- **Length:** 110mm (4.3in)
- **Sex differences:** No external distinguishing features but the female may be plumper, particularly when filled with eggs.
- **Aquarium breeding:** No reliable details, but occasional spawnings have been reported.

This fish seems to enjoy life tangled around plant roots and appears more often if kept in reasonable numbers. The rather long scientific name means 'thorn-eye' and refers to the spine beneath the eye; use a coarse net or plastic bag to trap it.

Below: **Acanthophthalmus kuhli**
This fish adopts a resting pose away from its usual haunt of plant roots.

Above: **Botia lohachata**
The striking colour pattern of this shy fish makes identification easy.

Botia lohachata

Pakistani Loach

- **Habitat:** Northern India
- **Length:** 120mm (4.7in)
- **Sex differences:** No noticeable features.
- **Aquarium breeding:** Not yet bred in the aquarium.

Unlike most Loaches, this genus is more 'fish-shaped', with the ventral surface flattened as befits a bottom-dwelling species. Most are fairly retiring fishes, preferring to come out late in the evening. They are very fond of worm foods and white worms in particular seem to attract them like a magnet no matter what time of day. They may make odd snapping noises.

Members of this species are a little intolerant of their own kind and of medications used in the tank.

Family: LORICARIIDAE

The members of this Family are known as Sucking Catfishes to fishkeepers but Mailed Catfishes to biologists. Their underslung mouths contain teeth that are used for rasping away algae, and the sucker-mouth enables the fish to anchor itself in position in fast water currents.

Above: **Hypostomus plecostomus**
This hardy and peaceful Catfish is active mainly during the night.

Hypostomus (Plecostomus) plecostomus

Sucker Catfish

- **Habitat:** South America
- **Length:** 150-200mm (6-8in)
- **Sex differences:** None known, but some Loricariids have bristly outgrowths on the head and thickened pectoral spines.
- **Aquarium breeding:** No yet achieved.

This gentle giant is a favourite for the aquarium, where it performs a very useful function in clearing the green algae. Once the algae has been cleared the fishkeeper must make up the deficiency of greenstuffs in the fish's diet with such foods as spinach, lettuce and rolled oats.

Family: MOCHOKIDAE
These African Catfishes include many species that are becoming very sought after among fishkeepers. They are not Mailed or Armoured like their South American counterparts and are often referred to as Naked Catfishes.

Synodontis nigriventris

Upside-down Catfish

- **Habitat:** Africa
- **Length:** 90mm (3.5in)
- **Sex differences:** No details.
- **Aquarium breeding:** Has spawned in captivity but the details are unknown.

A truly unusual fish, with the habit of swimming on its back. To help camouflage itself against the river-bed the belly is dark and the dorsal surface (now nearer to the aquarium floor) is paler in colour.

Synodontis species tend to hide away during the day and are active during the late evening.

Below: **Synodontis nigriventris**
This is the Upside-down Catfish in a typical pose. The upturned belly is mottled for protective camouflage.

Breeding Aquarium Fishes

Following the establishment of a community aquarium, the fishkeeper usually moves on to consider breeding his fishes, particularly encouraged by the appearance of young fishes (usually live-bearers) in his tank from time to time.

A large proportion of aquarium fishes may be bred in captivity and some fishes exhibit quite spectacular breeding habits and considerable parental care. Equally diverse are the actual methods of reproduction, which give each species its own particular challenge and excitement to the fishkeeper.

The fishkeeper has to bear the responsibility for providing the stimulus and opportunity for fishes to spawn and raise their young

successfully. This means selecting healthy breeding stock, bringing them to peak condition, setting up suitable breeding aquariums, supervising the spawning sequences and providing food for the young fishes. These are the duties necessary for a straightforward breeding sequence. If you are trying to produce a particular colour strain or finnage pattern, or trying to be the first to breed an acknowledged difficult species, then that is quite another story. This is one of the fascinations of fishkeeping: you set your own targets and, with a little determination and a lot of luck and enjoyment, you may eventually achieve them. It is a good idea to make notes of any attempt at difficult breeding, just in case it turns out right!

As the majority of fishes are egg-laying by nature, this form of reproduction must include a wide variety of methods to suit the different environmental conditions that are encountered by fishes in nature.

Unfortunately, the various methods of spawning do not coincide with our three levels of fish distribution within the aquarium, and so we must study fish breeding as a separate subject.

Egg-laying species range from the almost careless, spontaneous spawners to the ultimate in fry-care by the family-unit of the Cichlids. In between there are egg-burying, nest-building and mouth-brooding variants. In all cases, eggs released by the female are fertilized outside her body (although the point may be finely argued in the case of mouth-brooders, where the fertilization sometimes occurs in the female fish's mouth as she picks them up).

Egg-scatterers

Eggs released and fertilized in a spontaneous way are often safeguarded in nature by being swept away from the adult fishes by water currents. In the static conditions of an aquarium this cannot happen (although occasionally some may be caught in the filter box), and the eggs will probably be eaten by the adult fishes or others in the tank. When breeding, it is better to give the fishes a tank of their own, suitably furnished to their spawning requirements. To prevent the eggs from being eaten they must be separated from the spawning fishes in as short a time as possible. This can be achieved in a number of ways as illustrated in the diagrams on page 110.

Egg-buriers

Fishes from areas where streams dry out completely are able to ensure the continuation of the species by burying their eggs in the mud of the stream bed. In the aquarium a deep layer of peat moss makes an excellent substitute for mud, and the fertilized eggs can be stored in the semi-moist peat moss for several weeks before hatching is activated by immersion in water again. Eggs laid in plants (or

Above: *Egg-scattering fishes, such as these easy-to-breed Zebra Danios, should be prevented from eating their own eggs after spawning. They do not seem to realize that it is their future young (and fishkeeper's extra fish!) that they are eating.*

Above: *These Argentine Pearlfishes are taking the trouble to safeguard their eggs against desiccation when their stream dries up, by burying them in the stream-bed mud. In the aquarium a thick layer of peat moss is an ideal substitute. The eggs can be stored in this for several weeks.*

Above: *The fertilized eggs can be seen attached to this female* Oryzias melastigma. *They remain there for some time until they are brushed off onto aquarium plants as she swims about. The eggs are fertilized by the male during a nuptial embrace.*

Above: *Dwarf Cichlids often prefer to spawn in the privacy of a flowerpot on the tank floor, even if it means laying their eggs on the 'ceiling', as shown in this illustration. Usually these fishes are good parents, although sometimes one fish will drive the other away and care for the young alone.*

artificial spawning mops made of nylon wool) by some species – often referred to as 'switch-spawners' – may be collected by the hobbyist and kept in shallow containers floated in the main aquarium (or resting in a warm place) until hatching occurs.

One advantage of these delayable hatchings is that the eggs can be easily sent to other aquarists by post, thus ensuring more than adequate distribution throughout the entire fishkeeping world.

Egg-depositors

If egg-burying can be regarded as a step in the direction of parental care, the egg-depositing species take a much bigger step in the same direction, some species in this group being more complete and caring parents than others.

With their inbuilt desire to care for the eggs and fry, these fishes need little from the fishkeeper except a suitably furnished tank, depending on the type of spawning site preferred by the species. Egg-depositors require broad-leaved plants, pieces of flat slate or rock, and flowerpots or rocky caves for their spawning sites. The species *Copella arnoldi* requires a leaf overhanging the tank, as it lays its eggs out of water in order to safeguard them from danger; it may (alternatively) lay them on the cover glass. Additionally, the fishkeeper should keep an eye open for signs of pairing off between fishes in this group, as a self-selected pair usually make excellent parents. The majority of fishes in this group are from the Cichlid family; other egg-depositors include *Rasbora heteromorpha*, *Polycentrus schomburgki* and many *Corydoras* species, although these species do not practise parental care to the same extent.

Above: *The male Siamese Fighting Fish, in common with other Labyrinth Fishes, builds a nest of bubbles at the water surface in which the fertilized eggs are guarded and hatched. Only the male takes on the parental duty of raising the young fishes.*

Nest-builders

This group mainly consists of fishes in the Anabantidae group (Siamese Fighting Fish, Gouramies, etc) and one or two of the Callichthyidae (Armoured Catfishes). The nest is built by the male fish using bubbles of saliva, and often incorporating pieces of plants. Fertilized eggs are placed in the nest and guarded by the male, and the female is best removed once spawning is completed, as the male fish will often attack her.

Mouth-brooders

Many African Cichlids are mouth-brooders, although the Chocolate Gourami, *Sphaerichthys osphromenoides*, is also said to reproduce this way. Fertilization of the eggs may occur either before the

Below: *Siamese Fighting Fishes* (Betta splendens) *in a spawning embrace, during which the female releases between 400 and 500 eggs.*

Above: *Mouth-brooding fishes need no special breeding aquarium. The female hatches the fertilized eggs in her throat (taking no food during this two-week period) and the free-swimming young fishes also return there when danger threatens.*

Below: *A spawning pair of Dwarf Gouramies* (Colisa lalia). *The fertilized eggs are hatched in a deep bubble-nest and guarded by the male.*

female picks up the eggs from the spawning site, or as she does so encouraged by the imitation egg markings on the male fish's anal fin or spawning tassel. These fishes need no help from the fishkeeper; the female does not even take food during the incubation period, which usually lasts about two weeks.

Live-bearing fishes

Many female live-bearing fishes can store spermatozoa within their bodies, and successive litters can be produced from only one actual mating. Because of this, and of any male fish's persistent attentions, it is very difficult to keep the colour strain pure in any collection of assorted varieties of live-bearers. Members of the Goodeidae family require a mating for each litter; males of this group do not have a fully developed, rod-shaped gonopodium; only the first few rays of the anal fin are modified. The young are connected to the female by a placenta and these fishes most nearly approach the human mode of ante-natal gestation.

The live-bearer gestation period is usually around 30 days, with broods from 50 to over 200, depending on size of the fishes and the species.

Breeding techniques

The sequence of events for a successful spawning is as follows: selection of breeding stock, conditioning, spawning and fry care.

The selection of breeding stock begins with finding a true pair of fishes. This is easier in some cases than others, but observation of either the physical characteristics or the behaviour of the fishes will generally provide enough information.

Egg-laying species may be sexually distinguished by the fact that the male fish usually has brighter colours and more pointed or extended finnage, and is slimmer when viewed from above. The male's behaviour in showing off in front of a female or his repeated driving of her into a thicket of plants will also help to advertise his intentions. Cichlid fishes often pair off spontaneously, thus saving the fishkeeper the trouble of selection.

Live-bearing fishes are readily sexed, the male's modified anal fin making identification easy. Because of their readiness to spawn, live-bearers are popular subjects for line-breeding or for developing a new strain. Fishes for breeding are therefore selected for the qualities – such as size, colour or finnage – that they can add to the characteristics of the young fry.

To ensure strong healthy young, the adult fish should be in full health and at the peak of condition. They can be brought to this desired peak by feeding them amply with high-quality foods, and particularly with live foods. Additionally the sexes should be kept apart so that their excitement is heightened when they are re-united in the breeding aquarium.

The spawning of egg-layers should be supervised not only for the interest of watching the process but also to ensure that the female is not harmed by the over-zealous attentions of the male fish once spawning is complete. Occasionally, males will consider their mate not ready for spawning and will attack her rather than spawn with her. Sometimes it may be advisable to provide a conditioned male with an extra female fish or two so that his attention is divided at this time.

1

2

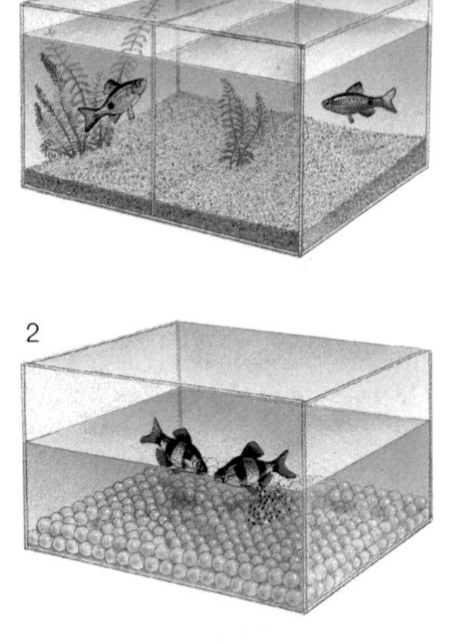

3

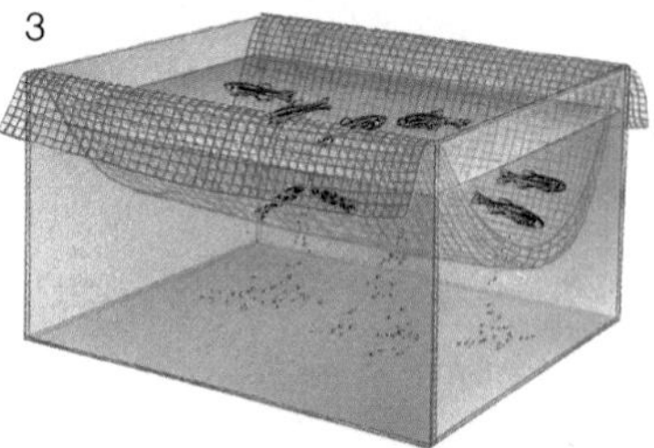

4

5

Above: *A spawning pair of Black Widow Tetras* (Gymnocorymbus ternetzi), *typical egg-scatterers.*
Left and below: *Fishes for breeding can be conditioned beforehand (1) by separating the sexes and feeding them very well. Egg-saving devices effective for egg-scattering fishes may be (2) a layer of marbles on the tank floor, (3) a layer of netting above which the fishes spawn, or (4) a clump of dense plants in which the eggs are hidden from the adult fishes. Egg-laying Toothcarps can be spawned over (5) a deep layer of peat moss, or (6) among nylon mops. Parental-caring fishes can be spawned in aquariums (7) furnished with suitable spawning sites according to the species' particular requirements.*

6

7

In general, both adult fishes of egg-scattering species should be removed after spawning, and it may help to shade the tank from direct light; some fish eggs are delicate and may be harmed by bright light. In some species the female only should be removed after spawning, the male assuming responsibility for any guarding of the fry, eg Gouramies. Where self-pairing has occurred both of the fishes should be given the opportunity to look after their fry, as in the case of the Cichlid fishes.

The fertilization of live-bearing fishes is a fleeting moment and the fishkeeper comes into the picture only when a female is obviously gravid (pregnant). This is apparent by an increase in belly size and the darkening of the area around the vent. Gravid females should not be moved towards the end of their gestation period, as this may cause premature births to occur. Confinement in breeding traps,

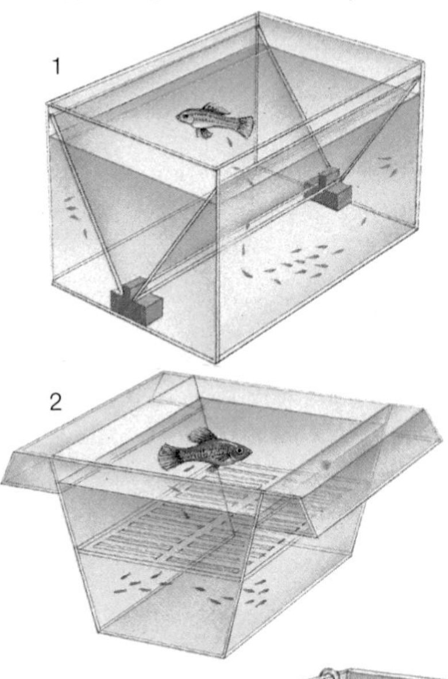

Left and below: *These artificial methods of saving newly born live-bearer fry are all variations on the same basic principle. Glass divisions (1), a separate floating breeding trap (2), or a simple netted area (3) will each prevent the female fish from devouring her young fry. Unfortunately, many gravid (pregnant) female live-bearer fishes are upset by being confined in breeding traps and may often give birth prematurely as a result. The floating breeding traps also keep the fishes in very warm surface layers of the water and this may account for adult female fish losses. It is preferable to provide the female with a separate, well-planted tank in which she can give birth undisturbed, recuperate from her ordeal and where the young can seek refuge from her attentions in the thick plants provided.*

together with the overheated water conditions where such traps are used, may also lead to premature births and even adult fish losses.

A separate, heavily planted tank is the best place in which the female can give birth, and the young fish immediately seek refuge in the dense plants away from their mother. Wherever possible the female fish should be given a few days' rest before being re-united with males once more in the community tank.

Care of the fry

Do not give food to the young fry until they are capable of taking it, or the unwanted food will pollute the aquarium water.

The young of egg-laying species remain immobile for the first few hours or days, feeding on the contents of their yolk-sac. When they become free-swimming, food should be readily available for them.

Live-bearer young will have ingested their yolk-sac while in their mother's body and will therefore be able to take food provided in the aquarium immediately after birth.

The growth rate of fry is obviously dependent upon the regular intake of food; it is important that the food can be easily taken by the fry, and that it is available whenever the fry wish to eat.

The food particles should be of the correct size for the fry; to this end, commercial brands of fry foods range from liquid to finely powdered from, and there are different nutritional recipes – eg live-bearers' or egg-layers' formulae – the former containing more vegetable material.

Many aquarists leave the fry-rearing tank dimly lit around the clock to ensure constant activity and feeding. With the young fry almost literally swimming in food, regular partial water changes will prevent pollution and also provide a stimulus to growth. A gradual increase in the water's depth over the first few weeks of the fry's life also helps to produce strong healthy fishes.

Reference to page 35 will provide hatching details of the brine shrimp eggs, *Artemia salina*, an ideal disease-free live food for young fishes. As the fry grow, the size of their food can be increased; screened live foods can be introduced and the fishes finally accustomed to their future full-sized diet. During the early weeks, any deformed or ailing fry should be discarded. Some smaller specimens may be late developers and worth keeping, but generally only the larger fishes should be reprieved.

Below: *The sex differences in these Mexican Sailfin Mollies* (Poecilia velifera) *are very clear. The male has a taller dorsal fin and an anal fin modified for fertilization.*

Index to Plants and Fishes

Page numbers in **bold** indicate major references, including accompanying photographs. Pages numbers in *italics* indicate captions to other illustrations. Text entries are shown in normal type.

Xiphophorus variatus (*Variatus Platy*)

Picture Credits

Artists
Copyright of the artwork illustrations on the pages following the artists' names is the property of Salamander Books Ltd.

Colour artwork
Bernard Robinson (Tudor Art Studios): 12, 13, 15, 17, 19, 20, 26, 29, 33, 34, 35, 41, 42, 43

Ross Wardle (Tudor Art Studios): 14

Brian Watson (Linden Artists): 24

Line artwork (Fish scale drawings): Sarah-Gay Wolfendale 49-103

Photographs
The publishers wish to thank the following photographers and agencies who have supplied photographs for this book. The photographs have been credited by page number and position on the page: (B) Bottom, (T) Top, (C) Centre, (BL) Bottom left etc.

John Allen Aquariums Ltd: 16(T)

Heather Angel/Biofotos: 83(B)

Bruce Coleman Ltd: 10-11, 50 (B, Hans Reinhard), 57 (Jane Burton), 82 (Jane Burton)

Eric Crichton © Salamander Books Ltd: 15, 18, 19, 21, 30-31(B), 32, 35, 37, 38, 39(T), 40-41

Jan-Eric Larsson: Title page, 25(T), 44, 51, 52, 54, 60-61, 68, 71(T), 73(T), 74(T), 99(B)

Malawi Tourist Office: 23

Arend van den Nieuwenhuizen: Endpapers, half-title page, copyright page, 27, 43, 45, 46-7, 48-9, 50(T), 53, 56, 58, 59, 65(B), 66, 67, 69, 71(B), 74(B), 75, 76, 77, 78, 79, 80-81(B), 81(T), 84-5(T), 86, 87, 88, 89, 90-91(T), 92-3, 94(B), 94-5(T), 96(B), 100-101(B), 101(T), 102, 103, 104-5, 107, 108, 109, 111, 113

Barry Pengilley: 28, 30(TL, TR), 31(T)

© Salamander Books Ltd: 39(B)

Mike Sandford: 70

W. A. Tomey: 22, 25(B), 55, 62-3, 64, 65(T), 72-3(B), 83(T), 84-5(B), 90(B), 96(T), 97, 98(T), 116

Willinger Bros., US: 16(B)

Editorial assistance
Copy-editing by Maureen Cartwright.

Acknowledgments
The publishers wish to thank the Aquarist (Chelsea) Ltd. for their help with location photography, and Interpet Ltd. for their help in preparing the book.

A superbly planted community aquarium